animals at peace

Alison Maddock

animals at peace

Harper & Row, Publishers
New York, Evanston,
San Francisco, London

Animal Life Series

Series editor: **Cathy Jarman**
Zoological consultant for this volume:
Michael Boorer

The chapter on animal co-operation was contributed by Christine McMullen

For information address Harper & Row, Publishers, Inc., 49 East 33rd Street, New York, N.Y. 10016.

FIRST U.S. EDITION
STANDARD BOOK NUMBER: 06-012728-7 (cloth)
STANDARD BOOK NUMBER: 06-012728-5 (paper)
LIBRARY OF CONGRESS CATALOG
CARD NUMBER: 71-185643

Contents

Editor's note

In this book, 'Animals at Peace', Alison Maddock presents a fascinating insight into animals as peaceful living beings. She discusses the day to day quiet routine of co-existing with fellow creatures in the family and group, as well as the astonishing co-operation between some totally different types of animals. Although animals may draw attention to themselves when they are fighting, most of the time their society is in fact well regulated and ordered – often more so than our own. In groups of animals where rivalries and disputes might be expected to cause strife, there is a code of social behaviour which ensures that life is, on the whole, peaceful. As well as co-operation in the family, odd partnerships occur between species with a common interest – whether it be defence, food-finding or comfort – like the hermit crab which tolerates and even encourages the presence of a sea anemone on its shell. In the same way, man forms relationships for mutual benefit with his domestic animals.

Engagingly written, this beautifully illustrated book is an excellent introduction to animal behaviour for children, students and adult non-specialists. It is complemented by a companion Pictureback, 'Animals at War' by Carolyn Barber, which discusses the warlike and aggressive aspects of an animal's behaviour.

How to be left alone

Glory be to God for dappled things –
For skies of couple-colour as a brinded cow;
For rose-moles all in stipple upon trout that swim;
Fresh-firecoal chestnut-falls; finches' wings.

from 'Pied Beauty' by Gerard Manley Hopkins

Day to day life is a continual struggle for animals. They have to find food, keep warm and out of danger. But there are moments of peace when every animal is free from these battles with nature. To achieve this peace much depends on the animal's ability to camouflage itself and keep out of harm's way.

What makes a peaceful scene in nature? Try to imagine and you will probably come up with a mental picture of a hot and sleepy summer's day, with butterflies flitting among the flowers and the air filled with the humming of industrious bees. Perhaps a family of stately swans is gliding across a nearby lake, and cattle are grazing in the meadows at the waterside. There is no fighting or unrest to disturb the mood of calm repose.

An over-sentimental description, perhaps, but one which contains most of the peaceful situations an animal is likely to meet in its life – freedom from hunger, cold and attack by hunters, and the comfort and protection of being part of a family or group. The ways of nature are harsh, and moments of restfulness free from struggle are rare. Many animal parents show a truly amazing devotion in the care and upbringing of their young, and this gives a certain degree of security. One of the requirements of the young is warmth. We tend to think of peace and warmth together because in temperate climates a warm day means a rest from the battle to keep out the cold. The animals have the same problem as people; they are constantly fighting against becoming too hot or too cold, too wet or too dry.

Getting enough to eat is another of the important struggles of the animal world, and for the more vulnerable species this means that there is constant danger of falling prey to a hungry predator – perhaps to man himself. In winter, conditions are hard and food is scarce, while in the summer and autumn months of plenty a good meal is much easier to come by. Gathering nectar is the summer work of the honeybees, whose regulated life ensures that food, like work, is shared between members of the hive. These insects have a part in our peaceful scene because they have perfected the co-operative life, living in large groups in which the common good takes precedence over all individual needs.

Avoiding the attentions of predators and competing with rivals for a mate or a home are both part of the age-old fight to stay alive. Undertones of this fight are always present and peaceful moments are merely interludes in the long struggle.

We have only to rouse the parental anger of the swans in our idealised picture to realise this. If the male swan thinks that someone is trespassing in his family home or molesting his offspring, he becomes a dangerous adversary capable of inflicting considerable injury. Only the cattle – domesticated, fed and protected by man – are more or less free from strife, though even they must compete with each other to some extent. Illustrating still further the delicate balance between peace and violence in the animal world is the fact that peace is often the hard-won result of previous battles. Thus the boundary of a male robin's domain is determined by the outcome of fights, or threatened fights, with his neighbours, but once he has established his territory he and his mate are able to raise a family there without too much interference from other robins.

So peace comes when an animal is engaged in activities which are directed neither at harming others nor at fighting to preserve itself from harm. This is a very individual kind of peace, and is probably truly present only when the animal sleeps, or among domestic animals and those protected in sanctuaries. But more important is the peaceful relationship between two or more members of the same species, which is vital for their survival. Not many animals are completely solitary, associating with other members of their species only for mating. Most have a form of social life at some time in their lives, when they must be able to adjust themselves to living with others in the family or group. Sometimes this involves overcoming an innate fear of too close an approach by another animal; this is done by taking part in ritualised displays such as the greeting ceremonies which strengthen the bonds between animals and ward off aggression between them.

Some animals have a highly developed social system in which large numbers live together and share in the activities of the group, a way of life best exemplified by the social bees and ants. Some of the most remarkable associations for mutual benefit, however, are between two or more animals of totally different species, like the mealtime cooperation between two animals with a common interest – the honey badger and the bird known as the honeyguide, or the mutual warning system of mixed herds of game and ostriches in Africa. The stories behind these partnerships are described later in this book.

Swans always seem to have an air of great serenity, though they are ready to defend their territory with force when necessary.

Keeping out of harm's way

The best way for a defenceless animal to keep out of harm's way is either not to be there when predators are around or to pretend not to be there. The first is attempted in several ways with varying degrees of success, by hiding away underground or adopting nocturnal habits, for example. The second is achieved by means of camouflage.

According to the evolution theory of Charles Darwin, animal species have become adapted to their surroundings and way of life because of the phenomenon of natural selection. Any new bodily or behavioural feature which gives some advantage to the animals that have it is 'selected for' in evolution and becomes more and more common over the course of generations. A feature is advantageous if it enables an individual to survive longer than its fellows and produce healthy offspring in greater numbers.

Evolution moulds the body into the shape best suited to meet the hardships of the environment, including the problem of avoiding enemies. An efficient escape method has a direct bearing on staying alive long enough to reproduce, and is often very highly evolved. The most important means of escape for one species may be the ability to run fast, for another to appear more frightening than it really is. But some of the most astonishing adaptations are to be seen in the animals which pretend to be some inedible object of no interest to predators. They do this by camouflaging themselves – as leaves, flowers, thorns, twigs, bark or even bird-droppings.

Nature is full of beautiful examples of camouflage. In

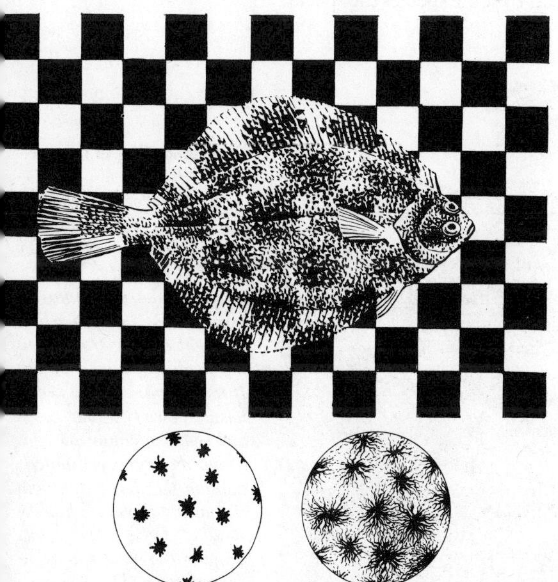

Opposite. *Asleep in the sun after a meal, this cat is warm, well-fed and at peace.*

Left. *Some flatfish have the ability to change colour to match their background. This trick of camouflage is sometimes so well developed that when a fish is placed on a chequerboard the pattern of squares soon becomes visible in the fish's markings. The appearance under the microscope is shown in the circles – when the pigment contained in special skin cells contracts* (left) *the skin colour is lightened, and when it expands* (right) *the animal's colour darkens.*

its simplest form camouflage implies merging with the background in such a way as to be invisible to enemies. It is often said that the animal is the same colour as its most usual background and so disappears when placed against it, but the story is not so straightforward. Human beings perceive animals not just as patches of colour but also as solid shapes, and it is likely that many other animals do this too. When light shines on an animal from above, its upper surface is bright and its under surface in shadow, and this betrays its shape. Many animals of all groups from mammals to insects overcome this by their countershading, that is, they are slightly lighter in colour on the underparts and this grades into darker hues on the back. This deceives the eye by cancelling out the shadows caused by sunlight.

Patterns of light and dark

We know that it is not mere accident that so many animals have lighter underparts, through the study of caterpillars which always hang upside-down under twigs. They are countershaded in the opposite direction – the under surface, which receives most light, is darker. These animals would immediately become visible if they changed their habits and started walking along the top of branches instead of underneath, so appropriate behaviour must evolve alongside structural adaptation. This is a general rule among camouflaged animals, which must both select a suitable background and remain absolutely motionless in order not to destroy the effect. In some species the special

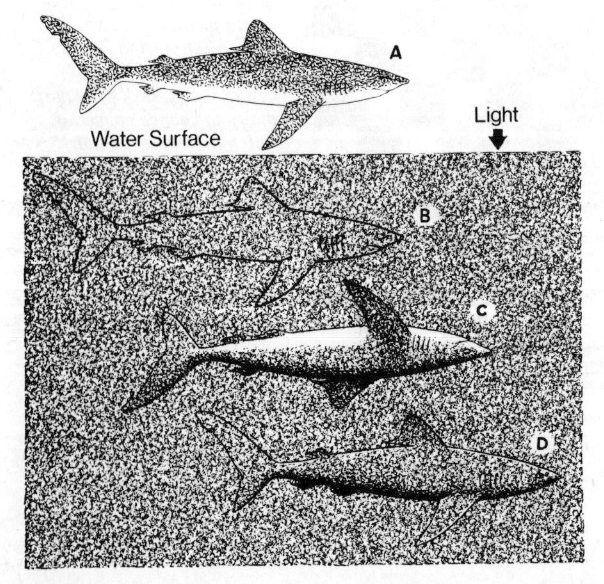

Countershading in the sea

A shark out of water (A) *can be seen to have a dark upper side and a pale under side, but when it is in the water the shark is almost invisible* (B), *because with the light coming from above the shading is cancelled out. If the animal were to swim upside-down* (C), *the difference in colouring would then be accentuated, making the shark stand out. A shark without countershading is also conspicuous because the underside is in deep shadow* (D).

Countershading on land

The caterpillar of the eyed hawk moth normally clings to the under side of twigs, and is darker on the underparts than on the upper. This countershading in reverse helps conceal the animal in its upside-down pose (top) *by giving it the appearance of uniform colouring. If the larva walks along the twig 'right side up'* (bottom), *then it immediately becomes much more likely to be spotted by a sharp-eyed predator, such as a hungry bird.*

behaviour is almost more important than the appearance of the body; when danger threatens, the blade-like shrimpfish shows the sharp edge of its body to the enemy so that it becomes almost invisible, sometimes enhancing the disappearing trick by hanging downwards among the spines of a sea urchin, which it resembles.

Animals which swim on the surface of water, such as the water shrew, are often darkly coloured on the back to make them less conspicuous to predators above, and silvery beneath to blend with the bright reflecting surface of the water as it appears from below. The dark and light areas are sharply divided, not grading into each other as in the countershaded animals just described.

A dark-coloured animal moving around on the snow is dangerously conspicuous, and so we find that many polar species have white fur or feathers, at least in the winter. The snowshoe hare is an example, being white in winter but having dark fur during the rest of the year when the snow has gone. The ptarmigan, a bird of the grouse family found in Scotland as well as more northerly regions, has three changes of plumage – brown and yellow in spring and summer, greyer in autumn and white in winter.

Colouring to suit the surroundings is found in all parts of the world; hares and other animals of open land are usually brownish with no distinct markings, whereas forest and tree dwellers, like the dappled deer, are often patterned in a way which makes them less visible in the patches of flickering light and shade. Birds of areas with sparse vegetation and sandy soil, such as the larks and other ground-nesting birds, tend to have brown plumage streaked with grey or buff and are extremely difficult to see when resting on the ground. There is some evidence that there are different races of lark, whose plumage colouring varies in accordance with that of the soil of their breeding place. In parts of Arabia, where patches of dark lava outcrop in the pale sand, there appear to be two races of desert lark, one of which has dark plumage and frequents the lava areas while the other has light plumage and is sand-dwelling.

Evolution at work

It is not often that there is an opportunity to demonstrate the effectiveness of newly developed camouflage, but a classic case which has been investigated by E. B. Ford and H. B. D. Kettlewell in England is that of the peppered moth, *Biston betularia*. The usual form of this insect is white peppered with black specks and it is well camouflaged when resting on lichen-covered tree trunks. Since about the middle of the nineteenth century, however, increasing numbers of a black or melanic variety have appeared. Originally a mutant or abnormal form, this type of moth must have become common because it had a greater chance of survival than the lighter variety when resting on the soot-blackened trees of areas which were affected by the Industrial Revolution. Moths of the 'normal' colouring are still the better camouflaged in areas where lichen flourishes, but they are easily picked out and eaten by

A ptarmigan cock in mottled summer plumage (left) *is well concealed on the rocky ground of its northern home, but when the snows come a whiter plumage is much better protection* (right).

ɔirds if they alight on blackened trees. This phenomenon s called industrial melanism and is often quoted as an ːxample of natural selection in action.

Examples of individual animals being able to change heir colour to match their background are not un-ommon, but the extent to which they can do so is often xaggerated. The flounder and other shallow-water flat-ish can alter their colour in response to light intensity, as an the chameleon and other reptiles, and this often esults in a better match with the background. The re-rrangement of pigment in the outer layers of the skin is esponsible for the colour change.

Some forms of camouflage help to break up the outline f an animal and distract the predator's attention from its ue shape; this is called disruptive coloration and is rought about by patterns of strongly contrasting colours. he European nightjar is almost invisible when it rests on eaths and bracken-covered slopes, not so much because harmonizes with the background but because its colours re broken up and the eye cannot see a coherent animal ape. This may also be the reason for the zebra's stripes nd the bright patterns of some fish. A predator such as a on may find itself unable to pick out an individual zebra om among a herd of dazzlingly striped animals, and the sitation this causes may give the endangered zebras st enough time to get away. Perhaps the stripes also aggerate the zebra's size, again confusing the hunter. nis method of using optical illusion to break up the tline has evolved in place of extreme modifications to the ody, because changing size and shape drastically tends to incompatible with the needs of movement and other portant activities.

On the other hand, however, some species have evolved zarre and beautiful appearances within the limitations the basic body shape which is part of their inheritance. ese are nature's real professionals in the camouflage t: past masters of the art of looking like something they not, and preferably something which is of no interest

Disguise and deception

Right. *Concealed frog; although insects probably display the widest range of disguises, there are some amphibians which are equally accomplished, like this horned frog from Malaysian forests.*

Top left. *A South America giant anteater giving her offspring a ride shows to perfection the disruptive coloration on the head and forelegs. A sharp banded pattern like this helps to break the animal's outline.*

Centre left. *The zebra's stripes are something of a puzzle. It was once thought that they camouflaged the animals in long grass wher the shadows cast are very sharp, but it seems more likely that they have a disruptive effect, so that a predator has difficulty in picking out the shape of a individual animal.*

Bottom left. *Not a plant but a living insect – the thorn tree moth's body is modified to resemble a double thorn, which helps protect it from enemies.*

to predators. Individuals thus disguised must live scattered, shunning the society of others of their kind, if the full benefit is to be obtained from their concealment. They must be the only unreal object in the collection of leaves, twigs, flowers or whatever it is that they are imitating.

Masters of disguise

It is difficult to select outstanding examples from the array of disguises to be found in nature – chiefly among the insects, but the most notable are probably the *Phyllium* leaf insects of Asia, whose bodies are entirely moulded in green leaf shapes, and those butterflies and other insects whose leaf disguise is so complete that their wings even bear the marks of veins and blemishes such as would be found on real leaves. The *Kallima* dead-leaf butterflies of southern Asia and Africa carry this camouflage on the underside of the wings only, so that when the brightly coloured insect alights and folds the wings over its back it immediately vanishes, to the great perplexity of predators. A 'tail' on the wings' rear margin touches the twig on which it is resting and looks like a leaf stalk.

Stick insects and the caterpillars of some moths are imitators of twigs, and like the leaf imitators must choose a suitable background and keep very still, or at most sway gently as if in the breeze, if they are not to be discovered. The same sort of camouflage is to be found in underwater animals, some of which have a normal body shape but disguise themselves. Some crabs cover their shells with seaweed, sponges and sea firs and replant them after moulting; freshwater caddis fly larvae make themselves tubular cases of any material which happens to be lying about – stalks, sand grains or bits of shell.

Opposite. *The caterpillar of the spurge hawk moth* (top left) *advertises its unpalatability by means of bright colours, but the angle shades moth* (centre left) *goes to the other extreme and disguises itself among dead leaves, where its mottled pattern and irregular wing shape are ideal for concealment. The prickly stick insect hides among bramble stems* (bottom left), *while disruptive colouring conceals the gecko on a tree trunk* (top right). *Of all these disguises none can match the ability of the* Kallima *butterfly to resemble a dead leaf (photographer's reconstruction bottom right).*
Overleaf. *Colouring to deceive: harmless butterflies which mimic poisonous species are quite common in Africa. A Nigerian subspecies of the golden Danaid butterfly* Danaus chrysippus (above), *distasteful to predators, is imitated by a palatable form of* Acrea encedon, *another Nigerian butterfly* (below).

Inevitably predators have become adapted to beat these animals at their own game. Built-in camouflage can equally well be used when lying in wait for a victim as when trying to avoid becoming one. Many crab spiders look exactly like the flowers in which they make their lair, and there are some fish which go so far as to have worm-like growths near the mouth to lure in the food. Nor is camouflage the perfect means of escape from enemies which track their prey by scent as well as sight, but it is certainly a widely used method of keeping out of trouble.

An unusual mimic – the rare British beetle Trichius fasciatus *imitates the markings of a bumblebee.*

Another way is to put on a terrifying appearance to scare off other animals. The idea is that very bright colours, for instance, warn that the animal is harmful, and for some species this is justified. The lurid black and yellow skin of the fire salamander advertises the fact that it is poisonous, but there are also creatures incapable of inflicting harm which adopt this trick. Especially curious are the cases of mimicry, in which an otherwise defenceless animal bears the colour markings of one which is poisonous or distasteful to birds and other predators. This is why hoverflies, which are often mistaken for wasps even by human beings, are banded yellow and black. The viceroy butterfly of North America, *Limenitis archippus*, is entirely palatable to birds, but it seems that owing to its excellent mimicry predators mistake it for the unpleasant-tasting monarch, *Danaus plexippus*. A bird which has never eaten the unpalatable species is not put off by its warning colours, but after a few unpleasant experiences it will recognise it and leave it alone. Experiments have shown that the harmless mimics are also left uneaten after this initial learning period, whereas other butterflies are taken as before. Many other examples of butterfly mimics are known, especially from Africa.

Staring eyes to terrify

Some butterflies and moths have eye-spot markings, consisting of coloured rings suggesting the light iris and dark pupil of the eyes of vertebrates – cats or owls for example. They serve to frighten attackers when they are suddenly exposed by moving the second pair of wings. Camouflaged moths without eye-spots remain still when approached by a searching bird, but those which, like the eyed hawk moth, have these special markings, open their wings and flash the staring eyes, often alarming the bird enough to drive it away completely. There are also some caterpillars which have eye-spots, and these are situated behind the head on a part of the body which can be puffed up or curved over to look like a much larger false head. When the body is also reared up and swayed from side to side, the appearance presented to predators must be very

This dull-looking Nigerian moth has a hidden defence mechanism in the suddenly revealed false eyes.

An adaptable moth

The peppered moth is a species which spends much of its time resting on tree trunks. When the trees of the industrial north became severely blackened by soot, this light-coloured moth began to suffer increased predation because it showed up so clearly on the dark background. Then a melanic variety, previously a rarity, began to increase in numbers as it had the advantage of being more likely to escape the notice of predators. Now the population of the peppered moth consists of both forms, the light one being more common where trees are lichen-covered and free from soot, and the dark one being the usual form in industrial areas. The pictures show how each is well camouflaged on the appropriate background.

reminiscent of a snake, and the message clearly stated is 'leave me alone'.

While not all animals are camouflaged by such elaborate means, many of them are at least coloured in such a way as to make them as inconspicuous as possible. Small, vulnerable species are often drably coloured and able to slip quietly through vegetation with the minimum of disturbance, like the tiny wren, whose movement in the hedgerow makes no more noise than a fluttering leaf. Such animals are usually safer living scattered and solitary than in large groups, and they tend to stay in undergrowth or trees, avoiding open land where they might be spotted by their enemies. The ability to keep absolutely motionless when danger threatens is of great importance, because this alone can make an animal almost invisible. Some mammals, such as lemurs, have shrill noisy calls, but their whereabouts are surprisingly difficult to detect because they have the ventriloquist's knack of concealing the source of the noise.

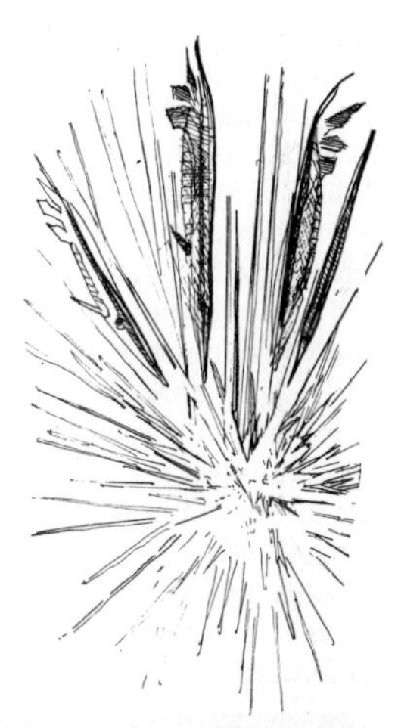

The seas contain a variety of strangely camouflaged creatures, and one of the strangest is the sea dragon of Australian waters (top). *The body is covered in leaf-like fronds which give this seahorse relative a strong resemblance to a piece of seaweed. Filefish* (centre) *have evolved the habit of hiding head down and swaying among eel grass, while shrimpfish* (bottom) *obtain their protection by sheltering among the spines of a sea urchin.*

Protected by darkness

Some animal species have transferred their main period of activity from the day to the night in order to remain hidden: they have become nocturnal, sleeping or remaining inactive during the day when the majority of large flesh-eating or carnivorous animals are about, and slinking out of hiding when the sun sets. This is no more than the habit of skulking about in dense undergrowth carried one step further. Both ways of life favour a more peaceful existence, but although the chances of falling prey to carnivores are reduced they are never eliminated.

The lemurs and their relatives are an interesting group in this respect. The true lemurs are distant cousins of the monkeys and are now found only on the island of Madagascar in the Indian Ocean, where they have developed into a variety of forms over the course of millions of years. In the earliest stages of their evolution they were to be found all over Africa and Asia, but with the rise of the more advanced monkeys their heyday came to an end. On the mainland there remained a few small, nocturnal species, now represented by the loris and the bushbaby, but on Madagascar they continued to thrive in the absence of competition and to develop into many new types. The reason for this was the isolation of the island – originally joined to Africa – during one of the periods of upheaval which characterise the earth's history, and the complete exclusion of the successful monkey group. Several of Madagascar's lemurs live in large groups and come out by day, in strong contrast with their shy relations on the mainland. Lorises and bushbabies are quiet and solitary, sharing with other nocturnal mammals a wide-eyed look which results from the great development of visual powers needed for seeing in the dark.

As an alternative to night-time activity, some animals build themselves a fortress for defence or burrow into the ground. A small hamster-like animal called the pocket gopher is one of many rodents which dig into the earth. It lives most of its life underground and so is free from the attentions of many enemies, but even its extensive burrow offers no escape from weasels and snakes. Termites are insects famous for the high mounds they build around their nests. Made of sand grains or wood cemented together with their own saliva and faeces, these may reach several feet in height and are so strong that only a hammer or pickaxe can shatter them. A few sharp-clawed animals are capable of breaking in – the aardvark for example, but the real enemies, against which even the termites' solid fortress is helpless, are bands of marauding ants.

For some animals there is safety in large numbers. Although it might be expected that a large flock or herd simply means more food for the predators, it seems that a dense mass of animals all moving together is confusing to an attacker, who is unable to concentrate on an individual unless one breaks free or lags behind. In addition, the many keen eyes, ears and noses of the herd have the advantage of making surprise less likely.

These are some of the ways in which weaker animals, and those not provided with weapons, avoid encounters with other animals likely to get the best of them in a straight confrontation. While insects and other lower animals rely to a great extent on camouflage, other methods are more common among the birds and mammals. But no method is ever foolproof; the economy of nature is based on the balance between a population's tendency to increase in numbers and its losses through disease and predation.

The wide-eyed charm of a young slender loris. Like its African relative, the potto, the loris is a quiet, slow-moving, nocturnal inhabitant of the forests.

A procession of parents

Sweet pretty fledgelings, perched on the rail arow,
Expectantly happy, where ye can watch below
Your parents a-hunting i' the meadow grasses
All the gay morning to feed you with flies:

from 'The Flycatchers' by Robert Bridges

Between birth and adulthood an animal is at the most critical and vulnerable stage of its life. Some offspring are ignored by their parents. Many, however, from spiders and earwigs to fishes and frogs get a surprising amount of attention. The peak is reached in the parental toil of the songbirds.

When spring comes to temperate lands the air is filled with the excitement of new life. The unfolding of leaves and flowers in response to the return of warm days coincides with renewed activity in the animal world and the birth of young animals of all kinds. Parent songbirds begin the incubation of their newly-laid eggs, and soon can be seen flying busily backwards and forwards with food for the hungry nestlings. In ponds and ditches tiny tadpoles are emerging from their eggs, wriggling to free themselves from the mass of jelly-like frogspawn, while on land baby mammals are taking their first steps under parental supervision. The renewal of life which each of these examples represents is the result of a male and female animal coming together to mate. In many animals this meeting signals the start of new and complex relationships between the members of the pair and between the parents and their young.

In some species the male and female animals meet only briefly for the purposes of mating, and do not live together as a pair for any length of time. In others, the association lasts for a much greater time – at least until the young are independent and sometimes longer. What brings the sexes together is a breeding urge which arises from the effect of inner changes such as hormone levels, which are caused by external changes such as the seasons. What keeps the sexes together after this urge has been satisfied is a new and less fully understood urge which expresses itself in the complicated patterns of parental behaviour.

It is usually assumed that all this is purely instinctive and involuntary, though some observers believe that there are animals which stay together because of something very like genuine affection as we know it. Konrad Lorenz, who has studied jackdaws carefully, believes that these birds form affectionate ties like those between human couples. When the male and female of a pair meet, they often indulge in ceremonies of recognition and greeting. Somehow this seems to ease the tension and relieve any distrust between them. The most attractive displays of this kind belong to the birds, and occur especially during the period of incubation, when the birds may co-operate in brooding the eggs. A bird approaching its mate does not behave in the same way as it would when displaying to a rival or enemy. Just as human beings smile to indicate their friendly intentions, so the bird signals by its actions 'I come in peace'. Its mate recognises this, responding with appropriate behaviour. The actions are not consciously performed or understood, but are automatic outward expressions of the bird's mood, resulting in an automatic or instinctive response from its partner. In the language of animals this adds up to a form of communication very necessary when two or more animals are to live together in peace.

The greeting displays of birds often take the form of feeding – the male may offer a small fish or other titbit to the female, very reminiscent of the 'peace offerings' in human families. Storks, on the other hand, throw back their heads and greet each other with a noisy clapping of bills. In birds which never leave the eggs unattended but take turns to sit, one cannot leave the nest until the other returns. When a bird ends his stint of incubating, the urge to brood has waned and the mood for food-getting takes over. After he has been absent feeding for some time, the brooding urge again begins to grow and he returns to the sitting bird to engage in a nest-relief ceremony. This helps the changeover from one mood to the other to coincide in the two birds. The herring gull utters a special call and presents nest material to its partner, which usually responds by leaving the eggs.

Jackdaws are among those birds which usually pair for life – in other words they maintain a complete and lasting 'marriage' through times when neither sexual activities nor parental duties require them to do so. Such species are often found to have an 'engagement' of some months before the actual mating takes place. Jackdaws and wild

A pair of lovebirds, living up to their name, indulge in a show of affectionate mutual preening.

geese choose their mates twelve months before they are sexually mature. To acknowledge her acceptance of the male strutting proudly before her, the female jackdaw crouches in front of him in a position which symbolises an invitation to mate. But the action has come to be used outside the mating season as a sign of greeting on any occasion when her mate returns after an absence. As in other animals, the ties between the pair are strengthened by special actions – the female begs for food using the postures of a baby bird, and in return for the shared meal she carefully grooms the head feathers of her mate.

Greeting is lcss obvious in many mammals, but as in birds there are many species in which male and female live in an amicable partnership, sharing in the defence and feeding of the young, for instance. The male may sometimes bring food to the female as she looks after their offspring. In many species, however, he plays no further part in family affairs after mating.

The first families

We are used to thinking of family life in terms of a father and mother and children of different ages. The parents form a permanent relationship and both help to feed, clothe, and generally protect the young. An older child is not discarded or driven away when a new one is born, but may help in the care of his brothers and sisters, who all learn from each other's experience. Such a well-developed family is very rare among animals, but a development towards it from the very lowly animals to the advanced warm-blooded birds and mammals can be traced. The most devoted bird or mammal mother may have to find a home, build a nest, keep the young clean and well-fed, warm them until they are able to regulate their own temperature, shade them from sun and rain, and defend them – sometimes from their own father.

The first hint of such care is seen in some of the arthropods or animals with jointed limbs – the group to which insects and spiders belong – where the mother stays with the young after they hatch. Sometimes it is actually the young which stay with the mother. She does not feed or warm them – they are cold-blooded and do not need to keep up a steady temperature in their body. Wolf spiders are so called because instead of spinning a web to ensnare flies, they run down their prey as wolves do. The female carries her eggs about in a cocoon made of silk which she attaches firmly to her abdomen. When the spiderlings hatch and emerge from the cocoon they climb aboard their mother's back and are transported wherever she goes. Although they must gain by her protection, she does not feed them and if one of the tiny offspring piled several

When the members of a pai of storks meet at the nest they greet each other with a characteristic display, bending the head back over the body and clapping the bill noisily in a signal of recognition and friendship.

Wolf spiders, like this one from Africa, are renowned for the way the female carries her newly-hatched young around on her back.

deep on her back should tumble off she does not go back to retrieve it. After about a week, the spiderlings go their separate ways.

Scorpions also transport their babies on their backs. They bring forth live young rather than eggs, born one at a time over a period of weeks. As a model of maternal care, however, they cannot match the humble earwig, the female of which behaves like a broody hen. She lays several dozen eggs in a chamber in the earth and gathers them into a pile, tending them by 'licking'. Up until this time the male has stayed with her, during the winter months, but now he leaves her to her brood, which hatches out in three or four weeks' time. A young earwig is born in a form known as a nymph, which is an intermediate form between the egg and the adult but is very like the adult in appearance. Many other insects go through a very distinct larval stage before reaching maturity, and there is usually no maternal care among these. The baby earwigs stay close to their mother and nestle under her, whilst they grow by moulting twice and eating the cast skins. Some insects, such as bees and ants, which live together in large numbers and are called social, show this sort of behaviour to an amazing degree. The bee community has been described as a superfamily, but it is very far removed from real family life and will be returned to in a later chapter.

Father minds the children

There is still a marked absence of family relationships in the lower vertebrates or animals with backbones, perhaps because so many of their actions are instinctive rather than learned. Animals with greater brain-power benefit more from being part of a family because they can learn by experience about the situations they will meet in adult life and about living with other animals of the same species.

Fish parents show us something new: a number of species leave all the hard work of baby-minding to the father, though most do not tend the eggs at all once they have been shed and fertilised. The seahorse is the little upright-swimming fish which looks like the knight in a chess set. The male keeps the eggs in a pouch on his belly, where they were laid by the female, and when they are born as miniature adults four to five weeks after laying, he seems to go through all the pains of labour. With jerking movements of his body, he forces as many as a hundred babies out of his pouch, one at a time with a rest between each effort. For actual care after birth, the three-spined stickleback is more renowned. He drives the female away after she has laid her eggs and he has fertilised them, and takes sole charge of the little nest which he has constructed for them. Now his actions cease to be dictated by the sexual urge and follow a new pattern of parental care, as changes in his motivation succeed each other rhythmically through the breeding season. His first job is to keep the eggs well ventilated, which he achieves by a fanning action of the fins, and this continues for the first day or so after hatching when the young remain in the nest. Soon they begin to move about and the father stops his fanning but remains diligently on guard, bringing back any over-adventurous young in his mouth. Gradually, however, he loses interest in them and ceases to retrieve stragglers as they become more and more active, though they seem to stick together of their own accord for a while.

A dried specimen of a male seahorse with the pouch cut open shows the developing young enclosed in the spongy tissue of his belly. They are born after five weeks.

Anyone who keeps tropical fish in an aquarium will be familiar with the many species of cichlid fish. Like sticklebacks, the parents fan their eggs, which are placed in a pit dug in the sand, but in this case both male and female co-operate in the task. Sometimes one gathers up the eggs in its mouth and transfers them to a new pit, while the other stands guard, and the fry are also guarded when they hatch. If one strays it is brought back in the mouth of the parent. This seemingly hazardous procedure is quite common in fishes. One cichlid, *Tilapia natalensis*, is called a mouth-brooder because the female holds the eggs and newly-hatched young in her mouth, and even after they have emerged she will quickly take them back in times of danger. Throughout this time some internal mechanism stops the female reacting to the young as if they were food and swallowing them – which would be a disastrous thing for the survival of the species if it happened.

In another cichlid, the jewelfish *Hemichromis bimaculatus*, colouring on the parents' body and a zigzag swimming movement seem to attract the young, so that they do not stray far from care. It has been found that the male and female of a pair recognise each other by their colours, and even after a period of separation reveal personal ties which last beyond the mating season. Such constancy appeals to us and shows that jewelfish come nearer than any other cold-blooded animal to the human idea of a family.

All the events and dramas of cichlid life take place in an area of the aquarium marked out by the male as his territory. The invisible boundaries of his domain result from sparring matches with his neighbours before spawning takes place. As an animal gets nearer to the centre of his territory – the nest – his boldness increases, so that he always wins fights, but as he chases his opponent away and gets further and further from his home ground and nearer to his neighbour's, his self-assurance wavers and he is liable to lose the next match. Chasing each other backwards and forwards the neighbours learn that there is a well-defined line between winning and losing. The territory is now established, and is maintained as an area safe from other fish until the young are old enough to lead independent lives.

A shoal of baby jewelfish is kept together because the young will always follow a parent swimming a zigzag course. When one parent relieves the other of its 'babysitting' duties, the fishes' behaviour makes sure the shoal stays together. The incoming parent swims straight into the group and starts zigzagging, while the off-duty parent darts away in a straight line.

'adpole travellers

rogs and snakes are not usually thought of as good arents. There are a few, however, which do not totally bandon their children. Britain's only native poisonous nake, the adder, has babies which, though independent birth, often stay near their mother for a short time. A ory which has been repeated many times since it first ppeared in a sixteenth century work is that of the mother dder swallowing her young. This has never been proved nd it may be that the young adders are in reality sheltering nderneath her body.

Frogs and toads pass through a larval stage before they e adult; unlike the snake babies which hatch or are born fully-formed snakes, their eggs hatch as tadpoles which n normally live only in water. A period of weeks or onths may be spent as a tadpole before the metamorph- sis or change into frog or toad is complete. In the curious urinam toad of South America this time is passed not in e water but, astonishingly, buried in the thick spongy in of the female's back. After the eggs – up to sixty of em – are fertilised, they are pressed into the skin by the

Strings of eggs from more than one female are carried around by the male midwife toad until hatching time.

Newborn adders are capable of independent existence, but they often seem to stay with the mother for a while.

The kokoi is a frog of the South American jungle. It is the male's task to carry the tadpoles on his back after they have hatched and to see that they are kept moist at all times.

male and become enclosed in little pockets, each with a lid, giving the female toad's back the appearance of a honey-comb. Young toads emerge about four months later.

Something similar is found in the female of the marsupial frog, *Gastrotheca marsupiatum*, another South American species, whose back bears a pouch for the young. The animal's name compares it with the pouch-bearing mammals, called marsupials. When it is time for the froglets to see the light of day for the first time, their mother lets them out by lifting a hind toe over her back and pulling apart the edges of the slit-like opening. She is not consciously acting for her offspring's good, any more than the wolf spider who transports her young or the cichlid who guards his brood. She is simply obeying an impulse to act in this way which arises when the froglets in the pouch are ready to make their own way in the world. There are other tropical frogs in which it is the male who carries the tadpoles on his back, attached by sucker-like lips, and who transports them to the water where they will complete their development.

But as far as family life is concerned, the most devoted frog or insect parent is outstripped by the warm-blooded birds and mammals. These have the most highly-developed systems of parental care to be found in the animal kingdom. Animals which do little in the way of caring for their children often produce many eggs and young because without parental protection their chances of survival are less. The more care given the smaller the family and, in many cases, the more helpless the young at birth. Individual care for a small number of offspring is the rule in most birds and mammals.

The world of a baby bird

Family life entails some very special relationships, formed with more than one purpose: between the parents the ties and interactions involve more than satisfaction of the exual urge alone, and between the parent and its young ctions take place with a variety of purposes, such as nsuring warmth, food and protection. These actions eplace others which would be normal outside family life,) that the young do not flee from their parents as they ould from other larger animals, nor do normal parents eact to the sight of their small children by eating them. omething about a hungry baby songbird causes its arent to feed it. That something is a gaping mouth retched upwards, often brightly coloured in the juvenile, nd uttering plaintive begging cries. But this is not enough n its own; the parent bird will probably not react in this ay unless it is at the stage of the breeding cycle when it ould normally be feeding young. Co-operation and ming are all-important; the needs of the offspring must e communicated to the parent, whose internal state must ake it likely to respond to the signals received. Evidently, hen one animal ceases to be a lone individual and starts ngling with other members of its species, its behaviour ecomes very much more complicated.

Newly hatched baby birds are described as being either dicolous or nidifugous, from the Latin meaning nest-ving or nest-leaving. Many songbirds are without athers, blind and almost completely helpless when they atch. For a long time they are unable to leave the nest and e fed continuously by their parents. These are nidicolous rds. Contrasting with them are the nidifugous gamebirds nd ducks. The downy nestlings are soon ready to leave e nest, especially to seek shelter if they hear the adults' arning cry, or to peck about for food.

The chicks of birds such as plovers, gulls and ducks are nidifugous – they are feathered and open-eyed when they hatch (top), *unlike the naked, blind nestlings of songbirds* (bottom) *which are called nidicolous.*

The blackbird and thrush are familiar songbirds. In ese species the female or hen is the only one to incubate e eggs, usually four in number, turning them from time time to make sure they are evenly warmed. Other species are the incubation between male and female; in doves e male takes the day shift and the female the night. After o weeks of incubating the blackbird nestlings hatch out, rawny youngsters with eyes closed and skin almost bare feathers. They seem to be forever hungry. A light tap the side of the nest sets them begging for food, their llow mouths gaping wide open. When they are blind, ey beg if they feel the sort of movement a parent return-g with food would make and they stretch straight up-ards. When their eyes open it is the *sight* of a moving ject which starts them begging. At first they still reach wards, however, only later beginning to stretch *towards* e movement.

Earthworms are the chief food of the chicks, and are ought by both male and female in the course of several ndred feeding flights a day. There is no special 'baby od' in this species, worms being an important part of the ult diet, but other birds have other arrangements. geon and dove infants are fed on a cheesy fluid called pigeon's milk for the first few days of their life. Rich in proteins and fat, this nourishing substance is made in the parent's crop and regurgitated when the chick pushes its head down the throat of its parent. Of course it is not true milk, which is found only in mammals. Many birds feed their chicks on regurgitated food, especially those which are absent on long fishing expeditions and return home only at long intervals. When a herring gull chick begs, it taps at a conspicuous red spot on the parent's yellow bill, which results in the adult regurgitating a meal. Even the newborn perform this instinctive act. In many birds of prey the father hunts for food for his family while the

As in other birds of prey, there is a division of labour between the kestrel pair when feeding the young. Th male usually hunts for the food, but he will first pass it to the female who then takes it to the nest.

Opposite. *Parent Caspian terns share the incubation of the eggs and later the care of the chicks.*
Overleaf; top left. *The first steps of a live-born viviparous lizard are watched by a solicitous parent. Other young are still surrounded by the egg membranes.*
Top right. *The conspicuous yellow gape of begging skylark chicks stimulates the adults to feed them.*
Bottom. *Mouthbrooding tilapia fish hatch their eggs out in the mouth and then carry the fry about in this way for some days. By the time they begin to swim for themselves, the fry are large enough to protrude through the parent's skin* (left). *Even after being released they may be taken up again for protection* (right).
Page 40. *Emperor penguins carry their single young about with them, tucked in between their legs.*

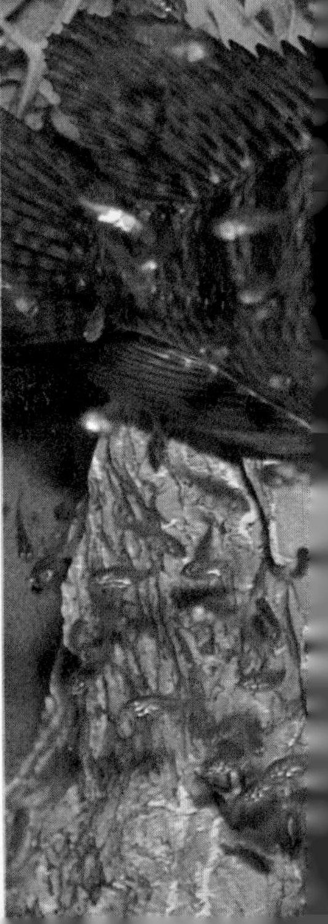

mother broods the offspring. Sometimes he does not feed the young directly but hands the food over to his mate first.

Hard work for the parents

Another parental duty is that of keeping the nest clean. First of all there are the broken eggshells to be disposed of, and the blackbird or thrush will take care that these are carried a little distance away before being dropped, so that their presence does not betray the nest's exact location to predators. Because ducklings and many other ground-nesting birds soon leave the nest anyway, their parents do not usually trouble to remove eggshells, nor are they so scrupulous in carrying out the blackbird's second cleaning job, that of removing droppings. In young songbirds these

The herring gull chick responds to a conspicuous red spot on the parent's bill by tapping at it, and this in turn causes the adult to regurgitate food for its young.

Keeping the nest clean: the adult song thrush removes the droppings of its young as they are produced, in compact faecal pellets.

are produced in neat capsules for easy disposal and are picked up in the beak and carried away by the hen. Birds' nests are the home of a host of tiny insects and mites, some of them parasites which are only too ready to attach themselves to a baby bird and suck its blood. Solicitous parents find and eat as many of these as possible.

With warmth and plenty of food the young blackbirds grow fast and in two weeks their nestling's down has given place to feathers. The fledglings leave the nest of their own accord but cannot yet fly and are still fed by the parents, which locate them by their cries. Not all parents of nidicolous young are so painstaking; petrels and shearwaters are seabirds which raise their single downy young in a burrow or crevice, brooding it at first but later visiting only at intervals to feed it. At length the chick becomes so fat that it exceeds its parents in size, and is then deserted by them to live off its stores of body fat until able to fly.

In autumn there is great excitement on ponds and rivers: it is pairing time in the duck community. The mallard drakes have newly regained their smart plumage and are

Fledgling turtle doves clamour for food from their parent, who gives them the cheesy 'pigeon's milk' regurgitated from the crop.

busy courting, displaying their finery to potential mates. Pairs stay together for a betrothal period until the spring, when mating takes place. The duck may build her nest in a variety of places, usually on the ground but sometimes in a tree hole as much as forty feet up – quite a problem for the ducklings when they are led to the water soon after hatching. The leap to the ground does not seem to harm them, however, and in one species of duck, the goldeneye of North America which always nests in trees, it is the normal state of affairs. Of the ten or twelve eggs laid by a mallard mother, few will reach adulthood, because there are many dangers facing the ducklings, especially from predators. Although the drake may not desert his family completely, he does not take on much responsibility towards them and it is the mother who shepherds her brood through their early life, supervising their feeding and flying wholeheartedly to their defence if need be.

Sharing the river with the ducks a pair of mute swans, which mate for life, is often to be seen. Sometimes the young cygnets ride on the back of one of their parents, tucked between raised feathers. Young grebes also hitch a ride in this way, and may be fed as they sit there, but a more unusual ride is reported to be given to woodcock babies. They are held between the legs and feet as this woodland bird flies through the air.

Babysitters Incorporated

Some birds are very community-minded in their child care. When young animals are grouped together and tended by a few adults, the group is called a crèche. Two or more families of eider ducks sometimes come together and guard all the ducklings regardless of parentage, but the most

Left. *When life gets too tiring for a baby black-necked swan there is always the chance of a comfortable ride on a parent's back.*
Below. *With heads down and bodies huddled together for warmth, a group of six-week-old emperor penguin chicks form a crèche.*
Opposite. *Mallard ducklings will follow their mother as soon as they are hatched, and respond to her alarmed quacking by hurrying immediately to her side.*

well-known of bird baby-minding groups are the crèches of some species of penguin. Each pair has its own nest and in most species both parents take turns at incubating and feeding, but when the chicks are partially grown they are placed in groups of twenty or more and left in the care of a few old birds while the rest of the adults go fishing.

Pairs of savannah cuckoos of the American tropics often combine to build a single bulky nest and each female lays four to seven eggs in it. All the parents co-operate in brooding and, later, feeding the chicks. There may be three broods raised in a year and it is said that the older chicks help their parents feed the younger ones. There are other birds which embark on the upbringing of a second brood before the first is independent. In the nightjar this is solved by the male staying with the young and the female sitting on the new clutch of eggs, while in the ringed plover male and female take turns with these tasks. Their behaviour must be very well synchronised to achieve this. Moorhens also have several broods in one year. They normally pair for life and share the building of a new nest for the second brood while the young of the first are still being fed. Older young have been seen helping with nest repairs and feeding their younger brothers and sisters.

A human mother can pick her child out in a nursery full of babies, even when they all look alike to everyone else, so it is interesting to see if birds can do the same. Herring gulls do not recognise their young at first; any young bird in the vicinity of the nest will be fed, and the rightful occupant will be ignored if it has strayed too far away. After several days the parents seem to learn to tell the chicks apart. Coots, like gulls, nest close together and so there is every opportunity for confusion over which chick is whose. Young coots return to the nest at night to be brooded, but tend to wander off during the day, and adult birds will feed any chick which is about the same size as their own. On the other hand, any 'wrong-sized' young are attacked. About two weeks after hatching, however, the parents begin to feed only their own young and drive out others, and at the same time the young cease to beg for food from any adult and concentrate on their parents.

A special kind of tie developes between the young of nidifugous birds and their parents. Born with a strong tendency to follow moving objects, the chicks instinctively trail after their parent as soon as they are active. Once the parent-object has been fixed upon, the chicks do not go running off after anything else and are said to be 'imprinted' on the parent. This word was coined by behaviourist Konrad Lorenz, who found that ducklings and goslings would adopt him as mother if he was the first moving object they saw. Unfortunately, the ducklings would not be convinced that he was mother until he crawled on all fours and made quacking noises – something which was a source of great bewilderment to passers-by.

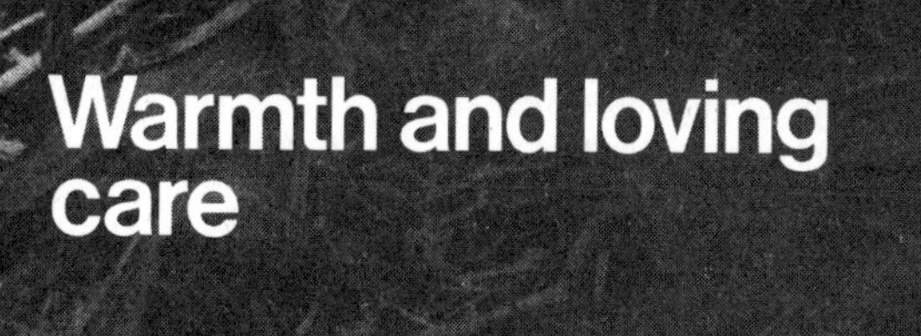

Warmth and loving care

Once upon a time there were four little Rabbits, and their names were – Flopsy, Mopsy, Cottontail and Peter.
They lived with their Mother in a sandbank underneath the root of a very big fir tree.

from 'Peter Rabbit' by Beatrix Potter

It is the mammals that take the greatest care of their babies. Although young deer and antelope can walk only hours after birth they still need attention. Baby rabbits are helpless for some weeks and the prematurely born marsupials are wholly dependent on their mothers for the first months of their lives.

Mammals are by definition warm-blooded animals which suckle their young. This means that there is contact of the very closest kind between a mother mammal and her offspring, which are at first entirely dependent on her for food in the form of milk. The physical and emotional bond which results is one of the strongest to exist between any two animals. Like young birds, baby mammals also rely on their parents for warmth and protection. Hand in hand with the great dependence of a young mammal on its parents goes a relatively long childhood, which allows for the full development of skills by learning.

At birth, the baby mammal may well be ready to walk and cope with its surroundings almost immediately. Baby deer and giraffe can walk about independently within hours of birth. On the other hand the young may be almost entirely helpless, like polar bear cubs, which require an especially long period of protection and care from one or both of the parents. This is comparable with the situation in baby birds dealt with in the previous chapter. Two mammal groups in which the great contrast is visible are the carnivores – foxes, bears and lions are good examples – and the hoofed mammals.

Male and female red foxes – dog and vixen – lead solitary lives except during the breeding season, when they stay together for the rearing of their cubs. When the litter is due, the vixen prepares a nursery burrow or earth containing a nest which she lines with some of her own fur. Her cubs, usually four of them, are born without fur and are blind and more or less completely dependent on her for food, warmth and protection. Meanwhile, the dog fox brings food from his foraging expeditions to give to the vixen, but he does not enter the nursery earth. Like all newborn mammals, the cubs are fed on their mother's milk, suckling at teats on the underside of her body as she lies in a position most comfortable for her and them. In this way mammal parents have an advantage over other species which feed their young, because the ready-made supply of milk does not necessitate a constant search for food such as many bird parents undertake. At ten days old the fox cubs' eyes open and they begin to be more active, but they do not emerge from the earth until they are nearly one month old, by which time they have been weaned and are eating solid food brought by the dog fox.

More lowly mammals, too, produce offspring in a similar state of unreadiness for life in the wide world. Among hedgehogs the female or sow has full responsibility for bringing up the young. The father is no longer around when the three to seven babies are born in summer. They are blind and possibly deaf as well, their ears being folded over. As for the prickles which will be their trademark in adult life, these are just two small patches of soft spines on the back at this stage. But they are not so helpless that they cannot find their mother to suckle and, growing quickly

A newborn rabbit (top) *is blind and almost hairless, but a newborn hare* (bottom) *has its eyes wide open and is well covered with fur.*

Mother love

Left. *Baby red foxes are blind and very dependent on their mother, who stays close beside them.*
Below. *Hedgehog youngsters taking their daily exercise with mother. The father has no part in family life.*

on the concentrated nourishment, they soon begin to look more like adult hedgehogs. The spines develop and harden, eyes and ears open, and at one month old the young are able to roll up into a ball just like mother does when frightened.

This is slow development indeed compared with the childhood of hoofed mammals such as sheep, cattle, antelopes and deer. Instead of a litter of several helpless children, the mother often produces only one offspring at birth. Equipped with a full coat of hair, with eyes open and ears pricked, the young animal can walk within hours. After twenty-four hours its legs are strong enough for it to run with the herd. Although it stays near the mother and is nourished by her milk, it requires little of the cosseting lavished on less well-developed young. The reason for such a difference is related to the contrasting feeding habits of the species. Many flesh-eating mammals have a home base where the young can be left as the adults hunt their prey, returning with a kill which forms a meal sufficient to last for some hours. Hoofed mammals such as antelopes, on the other hand, are grazers – they live in herds rather than family groups and are constantly on the move from one feeding ground to another. In addition, they are often in the position of being the hunted prey of carnivores such as lions. It is vital that the young should be able to keep up with a herd running to escape danger, and hoofed mammals are not designed to be able to carry their offspring on these occasions. Seals are in a somewhat similar position, though the young are more or less helpless for a few days after birth. A newborn fur seal pup has open eyes and a warm coat, and soon develops the ability to scamper about after its mother, joining her in a quick getaway if the need should arise. Nevertheless, the baby continues to be suckled for about three months.

Their stomachs swollen with two days' supply of milk, these baby tree shrews will be left in the nest by their absentee mother.

Absentee parents

Variations in the type of upbringing a young mammal receives are numerous. Rabbits and hares, which were originally thought to be rodents like squirrels and rats but are now considered to form a separate group, have a method quite different from rodents. It is sometimes called 'absenteeism' because the babies are left on their own for long periods. It is all the more interesting that this should be so because the state of development of newborn rabbits and hares contrasts strongly.

Rabbits are sociable animals whose homes are built close to each other in a system of interconnecting burrows. The pregnant female or doe, however, has her own territory, a short blind-ended burrow or stop-run away from the main system. Here she prepares a nest of hay and lines

Opposite. *Young deer and antelope can walk within hours of birth. As this three-day-old fallow deer takes a drink, its parent gently licks her young, a common way of establishing contact in mammals.*

it with her own fur in readiness for the litter of blind, deaf and nearly naked babies. As in many rodents and the hedgehog already mentioned, the father is an unreliable if not totally absent parent, and the doe must protect the young from him or he may kill them. Unlike the female fox, she does not stay in the stop with her offspring, but visits them once in twenty-four hours to suckle them for about three minutes, covering the entrance to the burrow when she leaves. After two or three weeks, when the eyes and ears are open, baby rabbits can run and make short excursions from the nest, and soon begin to take solid food. By this time another litter is on the way.

Baby brown hares, known as leverets, are strikingly different. They are born with eyes open and a full covering of fur, and can use their legs almost from birth. Each of the two to four young of one litter is placed by the mother in its own 'form', a concealed depression in long grass, which the female visits to suckle her offspring. One of the reasons a mother mammal stays near her young is to keep them warm, because in the early days of their life the mechanism which keeps the body temperature at a steady level is not working. It looks as though newborn hares and rabbits must be able to regulate their temperature better than other mammals, though the absentee method does seem to limit the size of each litter. In the warmer months a mother rabbit may have as many as eight babies at a time, but at other times the average is two or three.

There is another little animal which uses the absentee system, but because it is a shy forest inhabitant its habits were unknown until recently. The tree-shrew is a squirrel-like mammal from southeast Asia, living in pairs and building a separate nursery nest for the two or three babies. Weighing only half an ounce at birth, each tiny tree-shrew has eyes and ears tight shut and looks very vulnerable, but nevertheless it is able to survive a marked lack of parental care. Every forty-eight hours the mother visits the nest to suckle her young, which take in enough of the high-protein, high-fat milk to last them till the next feed, their stomachs becoming swollen with the two days' supply.

The importance of Mother

Suckling continues for a varying length of time in mammals. An elephant calf is not weaned until the end of its second year, in spite of being able to walk immediately after birth and to join the moving herd. Nor is it driven away when the next calf is born more than two years later. Both mother and child stand up to nurse, the baby curling back its trunk and sucking with the mouth. Most other baby mammals are weaned in a matter of weeks, like the

A female elephant ready to charge in defence of her offspring. Twins like these are rare in elephant families where single births are the rule.

In a series of experiments conducted at the University of Wisconsin by Harry F. Harlow, baby monkeys reared with artificial mothers were subjected to a frightening encounter with a noisy toy. Instead of the wire 'mother' on the left, which provided milk, the orphans sought the security of the towelling 'mother', which gave them comfort without food. Tests like these help to show the importance of a mother figure to a young mammal.

giraffe. Some, like pigs and cats, suckle when their mother is lying down, others when she is sitting upright. Dolphins are mammals which live in the sea, and so nursing presents some unusual problems for them. It takes place underwater, the mother squirting milk into her infant's mouth by means of muscular pressure on her milk glands, as the animals swim slowly along. The feed has to be a quick one, as the baby must surface every half a minute to breathe.

The monkeys and apes, man's closest relatives in the animal kingdom, cradle their young when they nurse them, just like human mothers. A very close relationship arises between a young mammal and its mother which is essential for normal development, especially in monkeys and man, where it plays a part in determining whether the child will grow up into a well-adjusted adult. Orphan monkeys in experiments adopted substitute mothers made out of wire and towelling, which they clung to for security and warmth.

Having the advantage over birds of a supply of mother's milk and prolonged development in the mother's body does not necessarily mean that young mammals become independent more quickly. Some are in fact born prematurely. These are the marsupials or pouch-bearers, most of which live in Australasia. The actual birth takes place at a very early stage of development when the young animal is a hardly recognisable lump of flesh, the size of a human thumbnail, with the mouth the most distinguishable feature. Female marsupials usually have their nipples covered by a bag or flap of skin, and it is to this pouch that the newborn animal crawls, attaching itself to a nipple and staying there until a much later stage in its development. The eight to eighteen young of the South American opossum would fit into the bowl of a spoon at birth. When they are old enough to emerge they ride on their mother's back. A kangaroo has only one young at a time, but that is quite enough for the parent to handle, because the growing 'joey' continues to ride in the pouch even when he is so large that his mother has difficulty in leaping. At the

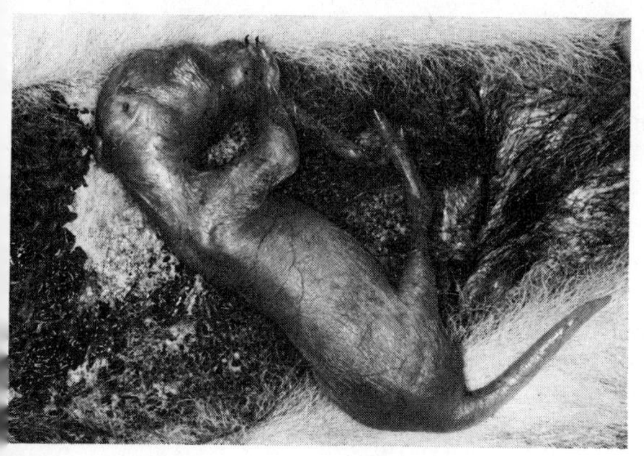

When a baby kangaroo is born after a 33-day gestation, it is scarcely more than ¾ in. long yet it manages to crawl over its mother's fur and into her pouch, using its sharp claws for grasping. There it attaches itself to one of the teats and begins to suckle. This tiny baby is only about 50 days old, and will remain in the pouch for over six months more. Even afterwards it will go on suckling for another six months before being weaned.

slightest sign of danger, the mother bends down and in hops her young one; if they are pursued, she is said to jettison the joey in a thicket and retrieve him later. The baby koala bear, another marsupial, remains in the pouch for six months before emerging to ride pick-a-back.

Holding the baby

Marsupials are not the only parents to carry their young; many other kinds of mammals can transport their young f the need arises. Often their aim in doing so is to escape langer, but some parents regularly carry the young until hey are able to walk independently. The hedgehog and nost carnivores pick up babies by the scruff of the neck nd carry them in the mouth, but sloth-bear cubs, like koalas and anteaters, ride pick-a-back. Even baby hippos have been seen on their mother's back, though this is unusual. Monkeys also ride in this way whilst their mother orages, though in the Old World species the very small nfant starts by clinging tightly to the fur on her underside, nd moves to her back at a later stage. Being very agile, monkey mothers are sometimes able to spare an arm to radle the infant as they climb through the trees, and apes early always do so.

A natural cradle is provided for the baby flying lemur: his southeast Asian mammal has no near relatives and is pecially noted for its ability to glide from tree to tree, sing a web of skin which extends from the sides of the hin right around the body, including the fingers, toes and il. Sometimes the mother hangs upside down in a tree ith the young flying lemur resting in the web, clinging to e fur by means of its teeth and clawed toes.

Horseshoe bats may carry their young in flight when ey are small, but later the mother hangs up her offspring the roosting place before setting off on a hunting expedi-on. But perhaps the most startling appearance presented by any animal family on the move is the shrews' caravan – a mother leading a long line of children in search of food, each holding on by its teeth to the fur at the base of the tail of the one in front. This behaviour is not typical of all shrews, however, and is not seen in the British species.

Good and bad mothers

In the wild, a mammalian mother is able to cope with the birth and care of her firstborn with an efficiency which is often astonishing. She seems to know exactly how to deal with the unfamiliar creature – how to cradle it the right way or to lie so that it can feed, how to protect it and to keep it clean. She often licks her new baby, which helps to establish contact and also stimulates the baby to evacuate the bowels. Her maternal behaviour is due partly to an inborn or instinctive knowledge, such as directs the behaviour of less advanced animals to a much greater extent. But in many mammals watching and learning from the more experienced females in the group is of extreme importance. A body of expertise and tradition is available

A female wild boar stands patiently as some of her striped young line up to suckle while others graze. Their baby coat probably helps to camouflage them in the forest during their vulnerable childhood.

Opposite. *The baby spectacled langur has an orange coat whose distinctive colour has the effect of eliciting parental behaviour in the adult monkeys.*
Overleaf. *Although little is known about the breeding of cheetahs in the wild, they have been successfully bred at Whipsnade Zoo in southern England. The cubs have a baby coat consisting of a grey mane which disappears after the tenth week.*
Page 60; top. *The mother raccoon of North America suckles her young in a sitting position.*
Bottom. *Baby African hedgehogs have bright pink heads and feet, and soft undeveloped prickles.*

to social mammals in their natural state which is often denied captive mothers. Sadly, complete success in breeding is frequently out of the reach of zoos because some female animals do not realise that the wet bundle they have produced is an infant to be cared for and fed. This is often because the mother's own social development is imperfect due to bottle-rearing. She may actually kill her baby, apparently in revenge for the pain it has given her in bearing it. If the infant is to survive, it must be removed from her company and reared artificially.

Basel Zoo in Switzerland is one place where even the most difficult animals have been persuaded to breed, and one of the star attractions there is Achilla, a female gorilla. In 1959 she bore the first gorilla to be born in captivity in Europe, a baby which was named Goma. Unfortunately Achilla did not know how to hold her infant properly to feed, and it had to be taken away to be cared for by human beings. To the great joy of her keepers, however, she discovered how to feed her next child, Jambo, and from then on she proved to be a successful and loving parent to her subsequent children, Migger and Quarta.

A baby mammal is often specially coloured to evoke parental behaviour in the adult. In some monkeys the baby coat contrasts strongly with the adult colouring, like the golden suit sported by a baby spectacled langur, whose parents are dressed in black and white. It seems likely that most mammal parents recognise their own children. Specific characteristics of the individual, such as appearance, the cries it utters and – of great importance in many mammals – smell, help a mother to know her child.

Because of the close ties between a mammalian mother and the infant she suckles, it is usually the females who bear the brunt of rearing the young. As in many birds, the males may mate with several females or the females with several males, and when the young are born blood relationships are not recognised. This is often the case in hoofed mammals such as deer. When this happens, fatherly care is not to be expected, though there are many other mammals in which the father does form part of the family and takes over some of the work. Often he is the food-getter of the family, but sometimes he performs duties normally carried out by the female. Two monkey relatives, the marmosets and the siamang, will often leave the cradling of the young to the father. In marmosets the young, usually twins, cling to his back and are taken by the female only at feeding time. The siamang is a gibbon which lives in family groups, inhabiting forests in southeast Asia. When the infant is weaned, the female seems to tire of her responsibilities and it is the male who cares for the growing animal during the important time of its life when it is learning to be independent.

It is not just the parents who may be called upon to lend a hand, however. A new baby creates quite a stir in the

The pride of Basel Zoo – Achilla and baby Quarta.

Baby transport

Top left. *The male cotton-headed tamarin, like other marmosets, carries his twin offspring round his neck rather like a shawl.*

Above. *After weaning her infant, the female siamang grows tired of its presence and leaves the job of carrying it to her mate.*

Top right. *The stoat is a Carnivore and, in the way of cats and bears, grips her young in the mouth when sh wants to transport them.*

Left. *A troublesome load: the young of the Virginian opossum, a North Americar marsupial, scramble onto mother's back for a ride.*

langur monkey population; other females cluster round the rightful mother and may be allowed to handle the infant. Observers have appropriately labelled this 'aunt behaviour'. Dolphins have aunts too. Two other females attend the mother at the moment of birth, and later a female may be seen to share the care of the baby with its mother. This is sharing on a small scale, but some species practise group care to a much greater extent. Like the penguins of the previous chapter, young giraffe and hippopotamus are grouped together in a crèche under the care of a few adults.

Play with a purpose

One of the most valuable aspects of being part of a family is having brothers and sisters. If they are of the same age they make good companions in play, and if they are older they may have experience which will be valuable to the younger members of the family. Learning from being one of a family is most important to mammals, which rely less on their instincts to direct their behaviour than birds.

Unusual foster mothers

Left. *Not so unusual as it might seem at first sight, this foster relationship between two animals on the farm arose when the sheep lost her lamb at birth and the piglet was orphaned at about the same time.*

Below left. *Even a hen may take to mothering baby mammals, though of course she cannot provide them with milk. These kittens have not lost their real mother but have transferred their affections only temporarily.*

The growth of mammals

The chart on the right compares the rates at which mammals mature and shows the relative duration of gestation (the period within the mother's body), suckling (the length of time until weaning) and childhood (the time taken to achieve sexual maturity). On the whole, the larger the animal the longer the gestation, except in the case of the kangaroo, born 'prematurely' at an early stage of development. The chart does not reveal all the different methods of parental care, as the giraffe and elephant can walk half an hour after birth, while a lion cub is much more helpless and goes on being taught and cared for by its parents for a long time after it is weaned. The column for man shows an interesting example of the effect of culture on the time of weaning. In some tribes it may occur as late as two years of age, but in western civilisation the desire to reduce the chores of parenthood has often brought it down to six months.

Play is one way of learning and practising skills which will be useful in adult life. Although young birds have been seen making playful hunting movements with inedible objects, or engaging in fights apparently for fun, play is not so important to them as to mammals like carnivores. These flesh-eaters have to learn all the tricks of the hunt: how to stalk, to fight, to pounce accurately, perhaps to climb trees, and so on. Kittens pounce at any moving object when they are young, and European badgers chase each other in a game of tag, but these seemingly useless activities help train them in muscular control and precision and teach them about their surroundings. Returning to the fox family described earlier, we find that when the cubs are old enough to come out of the burrow, their parents teach them to work for their food by playfully snatching

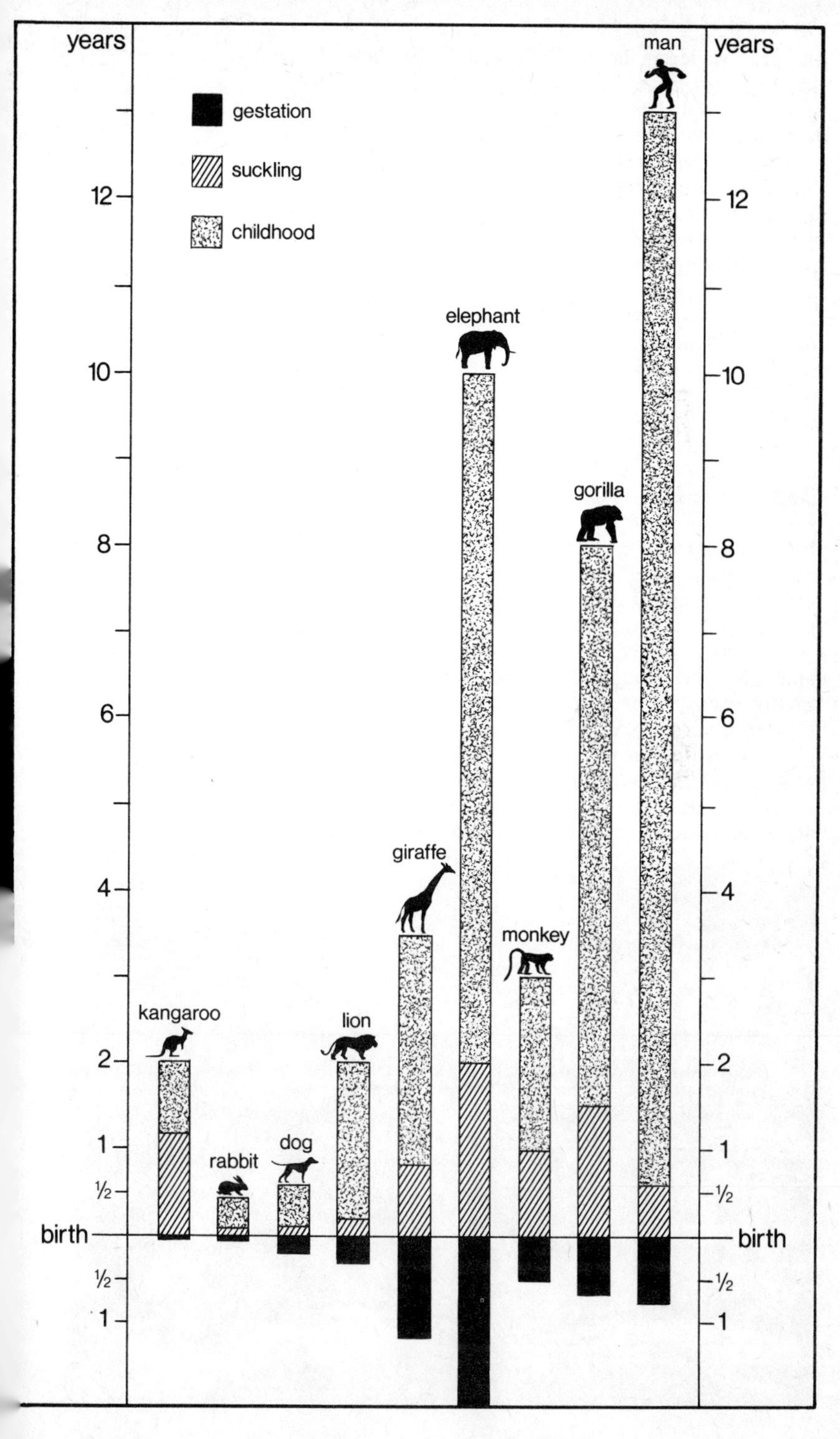
years
gestation
suckling
childhood
12
10
8
6
4
2
1
½
birth
½
1
kangaroo
rabbit
dog
lion
giraffe
elephant
monkey
gorilla
man
years
12
10
8
6
4
2
1
½
birth
½
1

the meal away from them until they catch it for themselves. Later the whole family goes hunting and eventually they split up to lead independent lives, but not before the cubs have learnt much from their parents.

Young hoofed mammals learn from their parents too, but the most important lesson to be learnt is how to run and leap to escape enemies, and this is practised by playing with other young animals of the herd rather than with the parents. Young fawns of red deer play at racing round a hillock, and lambs play 'king of the castle'.

When they are confined together in zoos, young animals of different species may play with each other, especially when one of the pair is that sociable animal, the goat. But this is not normal behaviour, and is rather like the touching relationships which arise when a mother animal is deprived of her own young. When this happens her normal rhythm of maternal behaviour is upset, and she may take to mothering almost any small animal she comes into contact with, even if it is a totally different species. Baboons have been known to adopt kittens and broody hens will also mother them until they are weaned. Sometimes this involves the coming together of two traditional enemies – like mother cats which adopt baby rabbits or mice.

All sorts of games are found among baby mammals – and sometimes among adults too – and not all can be simply explained in terms of learning for adult life. Chimpanzees sometimes adorn themselves with bits of plants or string as if they were dressing up, or mess around with paints to produce 'masterpieces' which sell for large sums of money. It is tempting to think that they are doing no more than thoroughly enjoying themselves.

The fishing lesson. Brown bear cubs experience one of life's adventures under parental supervision.

Nature asleep

**The north wind doth blow, and we shall have snow,
And what will the robin do then, poor thing?
He'll sit in a barn, to keep himself warm
And hide his head under his wing, poor thing.**

Anon.

Elephants stand to sleep, swifts roost in flight some thousands of feet up in the air, but the orang utan likes his comfort and builds a bed. Robins sometimes sing in their sleep but fish cannot even close their eyes to slumber. Ants take naps and bats sleep all winter, hibernating to survive the cold.

The sleep of animals is full of mystery and curious facts. It is not just a ceasing of activity, but a process which involves special behaviour and changes within the body and, at least in mammals and probably birds too, is necessary for life. We are not sure about the lower animals, but it seems likely that the sleep of mammals and birds shares many of the characteristics of our own. The harmful effect of a complete lack of sleep for more than a few days is well known in man and probably occurs in many animals. In sleep the body partially 'loses touch' with the surroundings as the heartbeat and breathing slow down and sensitivity decreases. There is often a drop in body temperature.

The animal often seeks out a particular sleeping or roosting place – perhaps up a tree or in a nest, though often on the ground or even on water – and it has certain postures which are always adopted when asleep. We may sleep sprawled out or curled up, on the back, front or side, and because we usually have the advantage of warm bed-clothes to protect us from the cold, it does not really

When the European red fox takes a rest, it curls up for warmth and the thick fur of brush makes a very comfortable pillow.

matter which of these positions we choose. But many animals must sleep hunched or curled up to prevent heat escaping from the body so helping them keep warm. It is rare to find any animal other than man asleep on his back, because this is the most defenceless and exposed attitude of the body. Both wild and domesticated cats and dogs are among the many species which usually sleep curled up, though they will readily stretch out in a warm place and doze.

There do appear to be periods of inactivity and reduced sensitivity in invertebrates and the lower vertebrates, such as fish, amphibians, reptiles and the social bees and ants, but whether or not this can be called sleep is uncertain. Butterflies and other insects may be found inactive at night, though this could be a direct response to the darkness and lowered temperature, whereas true sleep can occur at any time of the day or night. Some cold-blooded creatures become inactive during drought, and so do some shore-living forms when the tide is out.

Of course, sleep and wakefulness do not always follow a twenty-four hour cycle as they do in man, and in nocturnal animals their order is reversed. Some animals seem to sleep most of the time. The newborn human baby sleeps about five times a day, in comparison with nine or ten times for rats and mice and up to twenty times for rabbits. Cats have one long sleep in twenty-four hours but also several 'cat-naps', and guinea pigs seem to get their rest in very short bouts. They hardly ever stop eating and moving, but once in a while their eyes close and they seem to drop off for a few minutes. Whether animals dream or not is still a matter for investigation, but some – such as dogs – seem to start or grunt in their sleep; and some birds, like robins, sing. Quite often an animal rests during the day without actually sleeping, perhaps basking in the sun, the body stretched out to absorb as much heat as possible, or draped over the branch of a tree – the favourite relaxation of lions in some African national parks.

Below. *The leafy night-time retreat of the orang utan.* Overleaf. *A relaxed uakari monkey of South America sleeps draped on a branch.*

Sleeping postures

How do mammals sleep? The horse can close its eyes, relax its muscles and go to sleep standing on all four legs, but it may also lie down and rest on its side if very tired. Much bulkier animals, elephants have difficulty getting up once they are down, and must rock from side to side until they can roll back onto the feet. So the adults usually sleep standing, often with the weight off one leg, resting first one and then another. In a herd of elephants there are always some awake and on guard. When standing the elephant breathes at the normal rate, but when lying down – sometimes resting on a home-made pillow of vegetation –

it breathes at half this rate. The giraffe may seem to have a problem with its long neck, but it manages by sleeping on its feet and resting the head against a tree. When young it lies on the ground with legs folded and the neck twisted round so that it lies along the back; in this position the head fits neatly over the hind legs.

Many hoofed mammals sleep together in herds or flocks. Wild deer behave exactly like cattle, lying down and resting the head on the flank, but they do not seem to sleep very soundly as carnivores do. The digestive system of these ruminants requires them to chew the cud, so that they drowse rather than sleep deeply, and they are able to wake immediately they are approached because their excellent sense of smell warns them of danger. Man relies mainly on sound to wake him, and sometimes a considerable noise is needed before he will do so.

Sleeping together in a herd means that the individuals benefit from group protection. A herd of walruses or a group of hippos crowd so close together that they use other animals' bodies for pillows and are continually disturbing each other by their restlessness. Although Old World monkeys such as baboons live in groups too, they often seek the added security of the trees and sleep sitting up between forked branches. Tough, flat pads of skin on the rump help to keep the animal comfortable and balance it as it sleeps. Comfort is also the aim of the orang utan; at night he makes up a bed high in the trees by snapping off leafy branches and laying them across each other. As the nest rocks in the wind, the orang holds on tight by locking his fingers and toes onto the branches, a reaction so instinctive that one pampered orang placed in a human bed spent the night gripping the bedposts with hands and feet. Gorillas live in small groups and wander about feeding by day. At night they build sleeping platforms like the orang's but much lower down in the crotch of a tree, and the males position themselves nearer the ground as they are too heavy to climb far. They rest with the head on an arm and do not need to hang on to the branches.

Somewhere to sleep

Making a dramatic silhouette against the sky, a group of peacocks (right) *choose to roost in a tree, out of the way of ground predators. They sleep hunched up on a branch, but the hanging parrot of Asia* (below) *takes up a more unusual pose, head downwards. This behaviour is more familiar in bats, like the lesser horseshoe bat* (bottom), *which often roosts in caves. The grey seal* (bottom left) *is able to doze floating in the water.*

The wings of bats, the only mammals with the power of true flight, have often been likened to an umbrella, and the likeness is enhanced when they hang themselves up by the hind feet to sleep. Then the wings fold around the body like a closed umbrella or cloak, with the head hanging downwards. In the long-eared bat the ears are carefully folded and tucked away in the wing. Small parrots of the genus *Loriculus* also sleep in this upside-down way. Bats retire into hollow trees, caves and other dark corners during the day, sometimes roosting in large numbers with several species present or in traditional roosts used by the same colony for years. Male and female horseshoe bats sleep in separate places. When bats are asleep their body temperature falls drastically so that they enter a state of inactivity known as torpor.

Less is known about mammals which live in the sea, but walruses may sometimes sleep floating vertically in the water by inflating air sacs under the throat. Ships have several times reported collisions with sperm whales, and it is supposed that they were fast asleep.

Feathered repose

Birds are rather like bats in that they often like to roost in groups even though they may be unsociable by day. Bobwhite quails are North American partridge-like birds which sleep on the ground in a circle, their tails pointing towards the centre and their heads facing outwards. If a fox tries to sneak up on them, the birds suddenly explode into the air with a great whirr of wings, scattering in all directions to escape the bewildered predator. Thus their grouping seems to be primarily for defence. African mousebirds sleep in dense ball-like masses and this is probably a device for keeping warm. After a day spent feeding in family parties, coal tits also favour a crowded roost. At first there may be only one or two birds sitting on the branch, but soon another member of the group flies

The curious group roosting of bobwhite quails is a defensive system. Sleeping in a circle, they cannot be taken by surprise.

up and squeezes in between them, and before long the whole family is crammed into a small area, some of them even standing on the others' heads and forming a pyramid of birds. The most sociable of all birds at night-time are probably the starlings, which gather in flocks of hundreds of thousands in the winter and fly to cities and towns to roost on buildings, where they can become an extreme nuisance. Some of them may commute up to thirty miles every night to the roost.

Most birds choose somewhere safe to sleep because they are so vulnerable at this time. Some woodpeckers chisel out special roosting holes in trees. Sleeping afloat is the method used by many waterbirds, and they keep up a lazy paddling with one foot to remain in position. On land ducks often sleep standing on one leg, with the head turned round onto the back and the beak resting beneath the feathers between the base of the wing and the body – the position inaccurately known as the head 'tucked under the wing'. Storks, pheasants and some other birds do not use this posture but sleep with the head sunk on the breast between hunched shoulders. The need to keep warm influences the posture: birds may fluff up their feathers to create a quilt of air around them, and this insulating layer minimises heat loss. When perched on a branch or twig swaying in the wind, sleeping birds do not need to cling on because the toes lock onto the perch automatically as the legs relax, due to the tightening of a cord in the legs which pulls the toes together. On waking up, the bird stretches up its body and so loosens the clamped toes.

Beating the fuel shortage

Like bats, hummingbirds go into the state of inactivity and lowered temperature at night which is much more profound than sleep and is called torpidity. This is connected with the sort of life hummingbirds lead in the daytime. They fly with extremely rapid wingbeats which use up much

The fast beating of the wings of hummingbirds needs a constant supply of energy. At night they become torpid in order to save fuel.

energy, and fuel in the form of food must be taken in constantly. The tiny birds flit from flower to flower, sipping the nectar and also eating small insects. Components of the food, such as sugars, are 'burnt up' or oxidised within the cells in the energy-releasing process of respiration. At night food is not available and some means must be found of conserving energy. Torpidity is the answer, because the temperature of the body drops by as much as 22°C and the difference between this temperature and that of the surrounding cool air is greatly reduced. As a result less heat is lost, and the energy-consuming process of heat production is cut back considerably.

Baby swifts become torpid for the same reason in bad weather when the parents cannot feed them. The adults

Some fish definitely seem to sleep at night, and one of them is the parrotfish Scarus guacamaia. *It even goes so far as to secrete from the skin glands a loose mucous envelope, which surrounds the body like a sleeping bag. A hole at each end allows water to be drawn through and passed over the gills so that the breathing is not impeded. In the morning the fish breaks free from the mucus.*

Opposite. *South African ladybirds hibernate clustered together in a rock crevice. They choose a variety of nooks and crannies to pass the winter in, and some European species become a severe nuisance as they invade household equipment in search of shelter.*
Overleaf. *Herds of walruses sometimes heave up onto remote beaches to sleep and sunbathe, their bodies close together and their heads resting on each other's backs. Adults can also sleep floating vertically in calm water by inflating air sacs under the throat.*
Page 80. *Peacefully asleep among the daisies, a roe deer fawn adopts the curled-up position for comfort.*

themselves are able to pass the night on the wing, for while the nesting birds return to their holes, the non-breeders circle higher and higher into the sky and seem to vanish. This strongly suggests that they roost in flight, and radar has revealed flocks spending the night several thousands of feet up, probably sleeping briefly in between wing-flapping.

Birds and mammals can be seen to be asleep in a way we understand because their eyes are closed. Many cold-blooded animals, however, have no eyelids and so are unable to close the eyes. In their case it is difficult to be sure if they are sleeping. The goldfish in an aquarium seem to be always moving, but sometimes they can be seen resting with very little movement of the fins, and this may be sleep. A group of mackerel in a tank were observed for long periods but they never stopped their unending swim around the tank, though other fish such as wrasse are known to rest on their side or lie at the bottom of the water. Bedtime for baby jewelfish is indicated by their mother flashing her 'jewelled' fins to attract them to her. Then she leads them to the next hole while the father gathers up the stragglers in his mouth and blows them into the hole. By means of a reaction which contracts the swim bladder – the structure which keeps the babies buoyant – the young fish sink heavily to the bottom and remain motionless.

Even insects take naps sometimes. Ants seem to require sleep, and Julian Huxley wrote that they settle down for a few hours in a depression in the soil with their legs drawn up to the body. When they wake up all six legs and the head are stretched and shaken, and the jaws open in what looks surprisingly like a yawn.

Some butterflies, such as the small tortoiseshell, are able to survive the winter as adults by hibernating in a sheltered place – behind a picture, for example. Others overwinter as a chrysalis, the resting stage between caterpillar and butterfly.

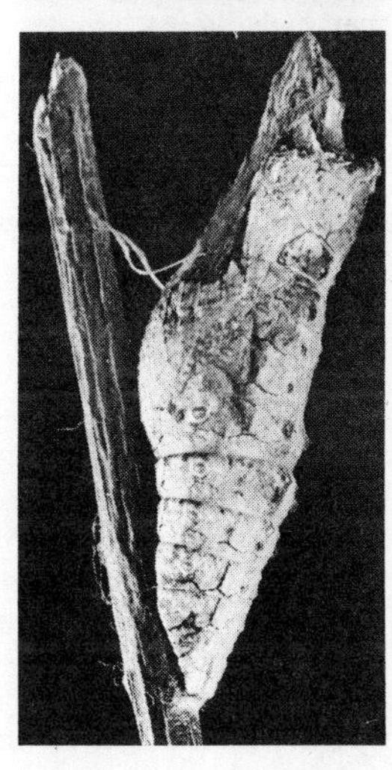

The deep sleep of winter

When winter comes to northern countries and high altitudes, a time of great hardship overtakes all forms of animal life. Food is scarce as many plants have died down or are covered by snow, and some animals solve the problem by migrating to warmer climates. In order to survive the cold and the threat of starvation, the animals which remain must build up their internal reserves of fat by copious eating in the summer and autumn months of plenty. Then they may, like foxes and deer, manage to live on this and the little amount of food to be had without taking any other special measures. Other animals may store up nuts and seeds, like the harvest mouse and some squirrels, and retire to a sheltered place to sleep lightly, waking often to feed and explore; or they may enter upon the long and deep winter sleep known as hibernation. In the less northerly latitudes, the occurrence of hibernation depends on the severity of the winter, and an animal such as the

When bats hibernate in caves, they are kept moist by the droplets of dew which condense on them. The same phenomenon is seen here on a glistening noctuid moth, hibernating in a damp mine.

badger, which hibernates in Scandinavia, does not do so in Britain, though it may enter winter quarters and become relatively inactive, sleeping intermittently.

Hibernation is far from being fully understood. Its onset is triggered not only by the coming of cold weather but also by the rhythm of changes within the animal's body. Both conditions are necessary and a pet hedgehog kept in a warm room will not retire to sleep, even if its 'internal clock' registers that it is time to do so. Many mammals, most reptiles and amphibians, and many invertebrates hibernate. The metabolism – chemical changes within the body – is greatly slowed down, so that the sleeping animal becomes inactive or torpid, having lost the powers of movement and feeling. The length and depth of this torpor vary with climate and species, but it is still not known why some species sleep and others do not. The deciding factor is not whether the animal lives on land or in water, nor whether it is cold-blooded or warm-blooded.

Many invertebrates – which are cold-blooded – simply get more and more sluggish as the temperature goes down and their metabolic rate decreases, finally reaching a state of 'cold anaesthesia' or dormancy, and resuming activity when the temperature rises again. Spiders seek shelter but do not hibernate; pond snails remain active beneath the ice, whereas land snails of the genus *Helix* stop eating in late autumn and creep into crevices, closing the shell opening with a tight lid and becoming inactive. There is no food for the honeybees when winter comes, so they cluster round the queen in the well-stocked hive and remain there until the spring flowers appear again. In other social bees and wasps all the members of the colony die except the queen, who hibernates. The tortoiseshell and brimstone butterflies and houseflies which appear early in spring have hibernated in corners such as behind curtains and doors in houses and sheds, but many adult butterflies die in autumn and their offspring pass the winter in the form of an egg, larva or chrysalis. Hibernation in insects may coincide with the phenomenon of 'diapause', a period of arrested development which takes place irrespective of the external conditions, so that the low metabolic rate and inactivity persist even in favourable conditions. Occurring at almost any stage of an insect's life history, diapause helps to ensure that the animal does not wake up on an unusually mild day in the middle of winter and use up all its food reserves.

Snakes, lizards, worms and newts of cold countries hibernate, clustered and twined together with members of their own species under logs or buried in a damp hole; this probably helps prevent freezing and reduces moisture loss. Garden tortoises also burrow into soft earth or under dead vegetation to hibernate. When the ice forms on ponds, frogs swim down and bury themselves in the thick mud at the bottom, or hide in soil, under stones or among grass roots. Toads dig into the ground. Eyes, mouth and nostrils are closed and enough oxygen from the soil dissolves in the body's moisture film and diffuses through the skin into the bloodstream to supply the animal's small needs. The body temperature follows the changes in the surrounding air temperature but is kept a degree or two above it by the metabolism. Freshwater fish may bury themselves in mud in winter, but hibernation is very rare in the sea because conditions there are so unchanging. However, it does seem that young plaice go into a state resembling hibernation, buried in the sand in shallow water.

Cold weather slow-down

These cold-blooded animals do not have the same problem as warm-blooded ones of keeping the body heat at a steadily high level. In true hibernation, mammals such as dormice and marmots abandon this temperature regulation and become virtually cold-blooded, remaining for weeks or months at a temperature little above that of their surroundings. They are inactive and take no food, their fat reserves providing the small amount of energy needed for breathing and circulation. In fact, experiments on hibernating marmots showed an apparently complete cessation of breathing, and a drastic slowing of the heartbeat so that the blood circulates very sluggishly. Digestion and excretion come to a stop, the blood composition alters, and the hibernating animal is often nearly completely insensitive.

There does not seem to be any rule about which members of a group hibernate: among the rodents beavers, voles, moles and lemmings do not, but dormice do. Often small animals which are most liable to lose heat and would not be able to keep warm in winter are affected. Hibernation usually starts when the temperature drops below 15°C, though some animals are active at lower temperatures, and most wake up at about 16°C. The waking is often quite rapid, and shivering helps to generate heat to send the body temperature up to normal. At the depth of hibernation, the body temperature may be as low as 4°C (the normal figure is a degree or two either side of 38°C), but if the animal is disturbed or in danger of freezing it usually wakes up because the heart beats faster and temperature control is resumed. Other reasons for waking in some sleepers are feeding and defaecation, though many live through the entire winter on their fat, emerging thin and weak in spring.

The nocturnal dormouse is the classical example of hibernation – indeed, its name derives from the Latin *dormire*, meaning to sleep. Even when not hibernating, it sleeps very heavily during the day, at a low temperature and seeming to be dead. As winter approaches, the daily sleeps get longer until they merge completely, and the animal dreams away the next six months rolled into a tight ball in a frost-free nest of twigs, moss and grass, buried snugly beneath grass or roots. Prepared by the

Escape from winter

Animals escape the cold of winter by several means. Many birds migrate to warmer climates, and gatherings of house martins on telegraph wires (below left) *are a common autumn sight in the countries of northern Europe. Other animals stay behind to face the cold and lack of food in their own way. Most small mammals collect a store of nuts or other food* (left) *and retire to a cosy nest. A few fall into the heavy sleep of hibernation, and pass the winter insensible. The common dormouse* (below) *has usually retired to its sleeping quarters by mid-October in Britain, curled into a ball with the chin resting on the belly, the feet folded round the muzzle and the tail wrapped neatly over the head. The body becomes so rigid and the sleep so deep that the dormouse can even be picked up and rolled over a table without waking.*

accumulation of much fat under the skin and sometimes the laying up of a store of nuts, the dormouse becomes very cold and almost ceases breathing, though it may wake from time to time to feed. The hedgehog's winter sleep is less regular than that of the dormouse, and the animal becomes less cold-blooded. Around the heart the temperature is normal but it decreases outwards towards the skin, the coldest parts being the tip of the nose and ears. A dark brown gland in the body called brown fat helps to keep the heat up – its cells are much more efficient at releasing heat than white fat cells. In the dry, moss-lined nest under leaves or in a hollow among tree roots, the hedgehog cannot store food as some animals do because its meal of earthworms would be unlikely to stay put.

All the British species of bat hibernate, fattening up beforehand in preparation. In other countries, such as the United States, the occurrence of hibernation depends on the severity of the local climate. The bats' sleep takes place under roofs, in tunnels and caves and other moist places where the danger of drying up is minimised, and clusters of roosting bats are often covered in droplets of dew. Some species sleep less deeply than others, waking up to fly about the roost or to hunt on warm days for insects which also wake up if the temperature rises above 8°C.

Sleeping curiosities

Not all sleeping mammals become cold-blooded. The bears, skunks and raccoons do not hibernate in the true sense, though they do eat a lot in autumn and get very fat before retiring to sleep in a hollow tree or cave. Pregnant polar bear and black bear mothers retire to winter quarters and produce their one, two or three young while in this 'twilight sleep', though they must rouse themselves sufficiently to bite through the baby's umbilical cord. The cubs are instinctively able to find the nipples and feed even

During the winter dark, the pregnant female polar bear prepares a lair where she will give birth to her cubs. The circular chamber is dug into the packed snow (shaded), on a hillside so that the fresh snow (unshaded) which falls during her sleep will not block the escape tunnel. In the lair the temperature is said to reach nearly 30°C, and the cubs are also kept warm by their mother's thick fur.

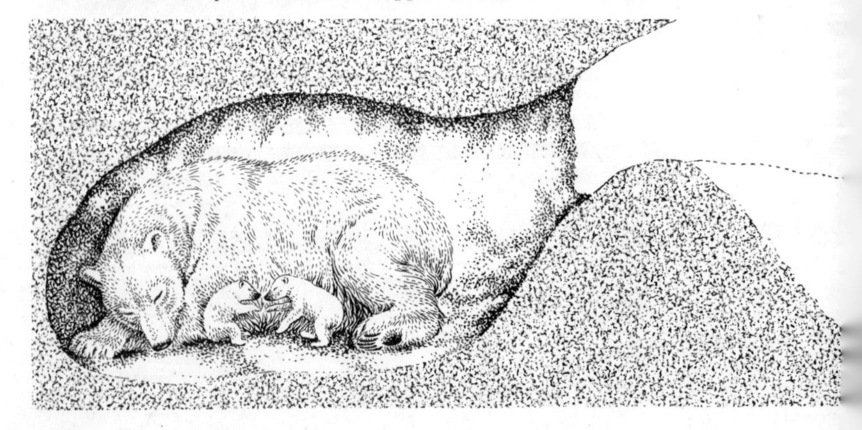

Travelling lungfish
The African lungfish (below) *is a lake-dwelling fish with the strange habit of burying itself in the mud during dry weather to avoid water loss. Armand and Michaela Denis found that this behaviour was very useful when they came to transport their small lungfish from Lake Victoria. The fish were placed in buckets of black mud, where they immediately wriggled out of sight, each cocooning itself in a watertight membrane. Nearly six months and many miles later, the mud having dried hard, the fish were retrieved* (left) *and returned safely to water.*

when the mother is asleep. They have grown considerably when after two months of alternate sleeping and suckling the family is led from the den on a warm day in spring.

But what of the other group of warm-blooded animals, the birds? It is very rare for them to hibernate, though for two thousand years tales were told of swallows and swifts passing the winter in hollow trees or even in masses buried in the mud at the bottom of ponds. Gilbert White, eighteenth-century author of *The Natural History of Selbourne*, believed this must be so in a few cases because they reappeared in the middle of winter on warm days, and because some fledglings were still in the nest in late September after their fellows had left the area. When the full details of migration became known the idea was dismissed, but from time to time there are reports of finding cold, torpid swallows and swifts in winter. The American nightjar or poorwhill has been definitely proved to hibernate on occasion, following the discovery of one in 1946 by Professor E. Jaeger which returned to the same spot for four successive winters. The body temperature was 22°C lower than normal. Twenty different kinds of normally migratory birds have been found hibernating, but it is not thought to be the usual way of surviving the winter.

Suspended animation

Hibernation has its counterpart in hot places in some parts of the world, and this is called aestivation or dormancy in dry periods. Mouse and dwarf lemurs in Madagascar are not very efficient at temperature control and become torpid for periods of several days in the dry season, living on fat stored at the base of the tail. Australia's spiny anteater behaves like a reptile in becoming dormant if the air temperature drops and food is scarce. Certain desert creatures survive drought by burrowing into the sand, like the water-holding frog of Australia which cocoons itself in a skin bag.

Some freshwater animals enter a state of suspended animation buried in mud if the pond or lake dries up. One of these is the lungfish, an air-breathing fish of African lakes and South American rivers, which gives out quantities of slime to form a watertight cocoon and lies doubled up with the head near a breathing tube in the mud. Scientists have found that the African lungfish can be transported in this state in cans of dried mud. It is alleged to be able to live up to four years in this way, though this must place a great strain on the body, since the fish lives by absorbing its own muscle, which is used as food. When the sleeper awakes it may find it has lost over an inch in length because of this curious 'cannibalistic' activity.

Home is the group

For so work the honey bees;
Creatures that, by a rule in nature, teach
The act of order to a peopled kingdom.

from King Henry the Fifth by William Shakespeare

The security of a social or family life is enjoyed by many animals not only for the advantages of defence and food-finding but also for companionship. Social units may be highly organised such as the bee superfamilies and the 'towns' of prairie dogs, or less ordered like herds of deer and antelope.

A family of animals, consisting of one or both parents and the young, may be thought of as a social unit. Some species take social life no further than this, while others flock or herd together in huge but temporary masses, perhaps for feeding or migration. Still others live in true communities, benefiting from being part of a close-knit group of all ages and sexes, and showing complexities of social behaviour quite unknown in the single family. These communities are the concern of this chapter, because they show how animals can band together to live in peace.

Group life has its beginnings when two or more families come together, like the eider ducks mentioned in an earlier chapter, or when the young stay with their parents into adulthood, as beaver kits do. Beavers mate for life, and their offspring stay with them for two years, forming a colony of up to twelve animals, living in a burrow or a 'lodge' made of sticks and mud. Damming a stream creates a suitable pond in which to build the lodge, which has underwater entrances and a central chamber above the water level, aired by a ventilating chimney. Although the construction work carried out by these animals seems to be skilfully planned, a large part of their activity is in reality instinctive. Nevertheless, in two years of living with their parents the young must learn a few tricks and refinements with which to vary their building. Learning is one key to

Most members of the cat family are not sociable and live rather solitary lives. The lion is an exception – a pride of lions consists of several females and their cubs, with immature males and an adult male.

the success of many social animals – in them its scope is greatly increased and the period over which it takes place is extended.

In human society there is a tendency for some people to be followers and others to be leaders. Animal groups are often equally in need of a leader, usually a big male who tolerates no insubordination from other males. A pride of lions contains several lionesses and juveniles but usually only one mature lion. As the male cubs grow up jealousies arise between them and the leader which are best resolved by them leaving the group, though the females are allowed to remain. Such a dictatorship is also found in the mountain gorilla, which has been studied in the hope that it might shed light on the origins of human behaviour. Small groups of gorillas wander about the African jungle, feeding on leaves and shoots and doing very little else. Unless attacked, they are very much more peaceful than the popular picture of them. The several females and young, with a few adult or subadult males, are under the rule of the biggest and most mature male, called the silverback male from his distinctive colouring. He is dominant over the other members of the group in the sense that he has first choice of resting sites and has the right of way along a narrow path, but he does not usually have to assert himself to get the respect due to him. His individual nature decides

Intent on their task, three hamadryas baboons indulge in a grooming session. This activity is very important in the lives of monkeys, because it helps to strengthen the social bonds which exist between them.

what the reaction of his group will be to another group encountered during the daily wanderings around the home range – friendly or otherwise.

Where social rank counts

While gorillas may be fairly near relatives of man, they do not lead such a regulated social life as another member of the same order, the savannah baboon of Africa south of the Sahara. The troops of twenty to eighty animals consist of several males together with more females and their young, and though mating is fairly indiscriminate, the big males extend their protection to any female with young. One male in particular is said to be dominant over all the others, because he always comes out on top in disputes, and he has a second-in-command who submits to no-one but his leader. In fact, all the males of the baboon society are arranged in what is called a hierarchy of dominance. Each animal is dominant over all the animals below it in rank but submits to those above. Position is determined by success in fights, so that when a group of strange animals are put together there is at first a period of restlessness and aggression. Once the rank-order is established, however, peace reigns because each animal

has learned his place and recognises exactly when to give in. A dominant male needs only to approach and perhaps threaten briefly and his subordinate will move out of the way, so that fighting is in reality reduced and the society stabilised. What we have described is called a straight-line system of dominance, first seen in farmyard hens and described as the peck-order. It is also the basis of jackdaw society. In fact, the situation in baboons is often complicat-ed by the fact that a low-ranking animal may dominate a high-ranking one when backed up by another individual.

The effect of the hierarchy shows when a dominant male akes first choice at food or a mate, but also when the roop is on the move. The dominant males keep to the entre, guarding females with small infants and determin-ing the group's movement by their behaviour, to which all he other baboons are attuned. Younger subordinate nales position themselves at the sides, front and rear and ct as sentries by alerting the troop with their alarm barks, n response to which the big males move forward to the ttack.

Hens in a farmyard are in an artificial situation, partly ontrolled by man. The hierarchy which is formed is not ke the baboons' or the jackdaws'. In these wild animals, ne lowest members are ignored by the highest, and the top nale is rarely challenged, showing aggression only to his nmediate inferiors. The unfortunate hen at the bottom f the peck-order, on the other hand, is often set upon and

When a troop of baboons is on the move, the dominant adult males do not lead, as might be expected, but accompany the females with young in the troop's centre. Other males walk at the front and warn their fellows of danger, and then the dominant males move forward to attack or defend. At the sides of the troop run the juvenile baboons.

The hens' peck-order is like the steps of a staircase. Each hen has a place in the order and gives in to those 'higher' than itself, while expecting submission from those below. When hens gather at a feeding place their turns at the food are determined by this order, an inferior bird being pecked at if she tries to queue-jump.

harshly treated by all above it. Moreover, this is a hierarchy of females, which normally play no part in baboon or jackdaw rank-orders. They may have a separate order or, more often, take on a status appropriate to their mate or their stage in the breeding cycle. A male's success in disputes may depend on the status of his mother in some monkey societies, but the chief causes are physical strength, age and self-assurance.

Knowing when to give in

Rank-orders are known to be the basis of society in several mammals, birds, reptiles and fish, including cattle, canaries, tortoises and sunfish. Although aggression is at the root of it all, fighting rarely does any damage because these social animals have what seems to be, in a loose sense, a conscience. When it looks like coming to blows, the fight is stopped by the subordinate animal making a gesture of submission – for example, crouching and withdrawing the head and making itself as small and inconspicuous as possible. Wolves present their most vulnerable part, the side of the neck, an action which completely blocks the attacking urge on the part of the other. In other words, he cannot bite even though he wants to. This

A dominance dispute between two wolves ends when one of them makes a gesture of inferiority to the other. In this instance he lowers his head and offers his neck to the snarling opponent, but sometimes the loser will roll over onto his back in complete submission.

instinctive restraint disappears if the victim now runs away, and the attacker will once more chase him.

The submissive gesture is designed to reduce aggression between rivals and mates, and sometimes evokes a sexual or parental response instead. The mating invitation of baboons known as presenting and the infantile begging of some birds serve to divert a superior's anger. On the contrary, non-social animals which are traditionally thought of as gentle – hares and doves – have not evolved gestures of submission, and when they are confined in captivity will often tear each other to pieces with no compunction whatsoever.

Obviously animals must recognise each other individually before the rank-order can succeed, and so memory comes into play. Lower forms such as ants and bees recognise that another insect is or is not a member of their colony, perhaps by chemical means, but do not know individuals. Cows, however, remember their position in the hierarchy after a separation of six to nine months. In these animals rank is determined solely by seniority; there is no competition for food, which is plentiful, or for anything else. Groups in which there is competition are more rigidly controlled and the best animal reaches the top, which is clearly beneficial to the society. Even in hens a rigid rank-order has a good effect – they eat more and lay more eggs. So the hierarchy system minimises physical damage, eases friction, encourages the fittest animals, and gives a stable, ordered framework for food-getting, mating, rearing the young and learning. Greater intelligence than their less sociable fellows is one of the noticeable results in baboons. The members of a group with a dominance hierarchy are in fact much more than a collection of independently-acting animals.

Disentangling the phenomenon of dominance from that of territory is not always easy. A society may contain both; for example a low-ranking pigeon, used to losing in squabbles on neutral ground, will defeat all comers when on his own ground or roosting place. Some monkeys with a hierarchy like the baboons' have a group territory, which all the animals defend, rather than individual

Opposite. *A large stinging jellyfish of worldwide range, the Portuguese man o'war is in reality a colony of up to a thousand polyps of three kinds, specialised for the tasks of stinging, feeding or reproduction.*
Overleaf; top left. *Baboons live in highly organised groups with a social hierarchy.* Bottom. *South American squirrel monkeys have more loose-knit troops.* Top right. *Suricates are mongooses of South Africa which live in colonies and enjoy sunbathing in groups, sitting bolt upright on their haunches.*
Page 100. *In partnership, a roan antelope and a pair of oxpeckers stoop to quench their thirst together.*

territories for each of the members. It has already been shown how the establishment of territory by fish means that fighting is reduced because one animal is dominant over all rivals in his domain. The large colonies of seabirds on rocky cliffs and islands are spaced out by the territorial activity of pairs of birds, and are less closely integrated than baboon or jackdaw society.

Underwater explorers have found that the territorial or 'capitalist' way of life and the anonymous 'communism' of shoals exist side by side in the fish of coral reefs. Plain-coloured or finely-patterned communist fish move about from place to place in shoals, while the capitalists, severely antisocial, stake out their territory by a piece of coral or weed and defend it fiercely against others of their own species. Their brilliant colours and patterns make them conspicuous to predators but, more importantly, proclaim the territory and warn off rivals. Who is to say which is the better way of life? Both have proved successful, and there is room for all sorts amid the teeming life of the reef.

Underground townships

Baboons do not have a particular territory, but they stick to a home range as they wander about foraging for food. Other mammals have a fixed home in a burrow or nest. Prairie dogs are ground-living rodents of North America, whose claim to fame is that they live in huge underground 'towns', networks of burrows covering many acres and

A touching scene, but full of meaning in prairie dog language. These rodents recognise each other as friend or foe by means of the kiss, an important act in their society of well-ordered relationships.

containing thousands of animals. Larger towns are divided by natural features of the landscape into wards, but there is visual and vocal communication between wards. Within the wards are the basic units of this complex society, the coteries, each of less than one acre and derived from a family group of up to thirty animals. The adult male, sometimes with a subordinate, and his several females with young defend the coterie's territory, advertising their rights by rearing up on the hind legs and calling. When one prairie dog meets another, they drop onto the belly and crawl forward to exchange a kiss. If they are friends, they may proceed to groom each other – a favourite occupation – but if one is an intruder from another coterie he is chased off. These highly-organised animals even have overspill towns and suburbs to make sure that the population is well spaced out and does not compete too much for grazing. Young males become restless in the breeding season and may move away to dig their own burrows. Even an adult female may feel things are getting too crowded as the children grow up, but instead of driving them out she will leave them with the ready-made home and find new quarters herself.

Rabbit warrens do not extend over so large an area as prairie dog towns, but they are often a good size and well populated, as rabbits are very gregarious by nature. The excavations include bolt-runs and emergency exits, living chambers and blind-ended tunnels. A male has his little territory within the warren, centred around his females, and he has a rank in the community depending on seniority, size and fighting ability. Living close together, normally an advantage, proved the downfall of many rabbits in Britain in the 1950's when myxomatosis seared through the population, the infection spreading rapidly from one member of the community to another.

In coatis, South American raccoon-like mammals, the adult males are kept out of the communal bands, which consist of about twenty females and young. Only in the breeding season are males allowed to join the group; at other times the light-hearted band of females sleeps and forages together, relatively free from dominance disputes and aggression. A group patrolling its home range occasionally encounters other coatis in the same area, and there is some argument but little fighting.

The coati's society can be called matriarchal, because it is the females which call the tune. This is now known to be true of the sociable hippopotamus too: 'hen-pecked' males are obliged to keep their distance from the central crèche of youngsters jointly cared for by the adult females. This crèche is situated on a sand bar or bank by the river, and is surrounded by the refuges of the males – each a domain of a few square yards occupied by one animal, which has fought for a place or rank in the 'queue' to be near the females. The latter choose their mates and also pay social calls on them outside the breeding season, but a male which dares to return the visit must behave with extreme deference. Hippo mothers have rather the same tyrannical attitude towards their young, which must at all times follow their parent obediently and closely. Not surprisingly, young males are driven out of the crèche as they mature, to refuges on the outskirts of the group, and they gradually work their way inwards as their seniority increases, until they are among the privileged few allowed to mate with the matriarchs. In this species males are dispensable for a large part of the year as the females have the main responsibility for rearing the young, and so this society ruled by females probably serves to maintain order and peace by keeping the aggressive males at a distance.

Communist flocks and herds

Very large flocks, herds and schools of animals represent a less ordered society, in which there is no one place called home, and the main value of the grouping lies in protection from enemies and communal food-finding. In fact, the way of life is a sort of anonymous communism with little individuality but also little fighting except at mating time. With this less stimulating background the baboons' feats

Placidly sunning themselves on a sandbank, a group of hippos do not betray the extraordinary dominance of the females, whereby the males are forced to live in refuges at a distance from the central group of females and their young.

of intelligence and learning are not reached – or needed.

But the members of a herd are not simply thrown together *en masse*; they act together in a synchronised way, feeding and breeding at the same time. Thus all the seagulls in a colony nest at the same time and are able to take part in group defence against attacks on the eggs and chicks by predators. The familiar chirping of house sparrows is the means by which members of a flock communicate and stay together. They feed, dustbathe and roost together and fly off in unison when the alarm call is uttered by any one of them. Flocks of Canada geese, which graze on grasslands, are often described as posting sentries following a regular rota so that each bird takes its turn. Probably there is no such deliberate process, but any bird which happens to find itself on the edge of the flock will raise the alarm if it sees danger.

Animals in flocks and herds do not usually have a very large repertoire of calls, because there is a limit to the kinds of information they need to communicate. Social monkeys have a far greater range of sounds, and so do the advanced and intelligent dolphins, though there is nothing approaching language as it occurs in humans. Only recently have the noises of dolphins – creaks, whistles and grunts made by forcing air out of the blow-hole – been analysed and found to be a form of conversation between members of the school. All ages and both sexes live together harmoniously with a peck-order based on size, and if one dolphin is injured others will buoy it up so that it can breathe.

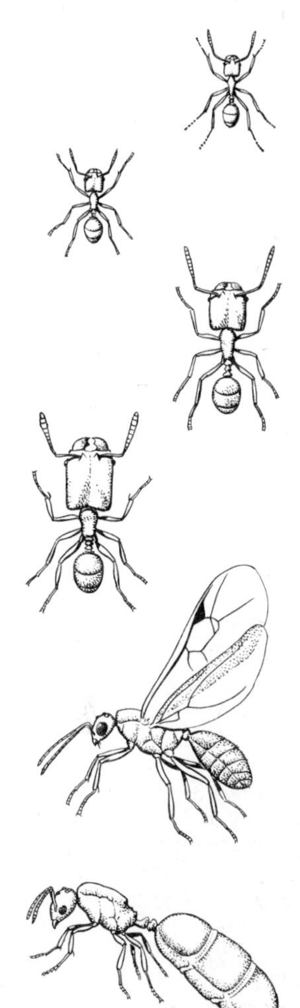

Above. *Ant communities show a division of labour which is matched by the different forms of the members. In the Argentine ant there are winged males, several sizes of wingless workers and soldiers, and a large queen with a swollen abdomen.*

Opposite. *Two very sociable species, the dolphin and the gregarious herring gull.*

The superfamilies

Social life takes a number of forms other than the anonymous herd or the territorial group with close relationships. There are associations of lower animals which are far more intimate and controlled than anything practised by birds or mammals. Instinctive behaviour allows the high level of organisation of the large bee or ant co-operatives, in which no one individual has all the benefits from the community's work but all share equally, dividing the activities between them. This division of labour is so pronounced as to have resulted in the evolution of different physical forms for carrying out the different duties. Individual ants or bees are rather stupid, but the whole swarm acting together has the appearance of a super-intelligence.

Most European honeybees are domesticated and live in man-made hives, though some occupy hollow trees. All the bees of one colony are related, being the offspring of one mother, the queen. In summer there are three types or castes of bee in the hive; the egg-laying queen is a female with a somewhat swollen abdomen; her rôle is to produce up to fifteen hundred eggs per day. She may live from two to five years, controlling and co-ordinating the behaviour of the rest of the hive by the presence or absence of secretions from her body. In termites, commonly called white ants but actually distant relatives of cockroaches, the queen's abdomen is grossly out of proportion to the rest of her body, it is so swollen with eggs. Her lifespan may be fifteen years or more.

The other occupants of the honeybee hive are a few hundred drones and twenty to eighty thousand workers, each with a life-expectancy of four to six weeks. Of the queen's eggs which are fertilised most develop into sterile females – the workers. Sometimes, for example when the old queen meets with an accident, a fertile female – a new queen – develops. To achieve this a larva is given a specially rich diet of 'royal jelly'. Unfertilised eggs develop into the males or drones, whose sole purpose is to mate with the queen; they do not work, and are fed by the workers until, when autumn comes, they are turned out of the hive to die. The individual does not count in these ruthless insect societies; the good of the colony is the end to which all behaviour is directed.

Workers have a variety of tasks in the community, depending on its needs at any one time. In the first days of their adult life, their time is spent as nursemaids to the eggs and developing larvae, and tending the queen. They produce food for their charges by regurgitation and by secretion of jelly from the salivary glands. At the end of ten days there is a changeover to work on the upkeep of the hive – building and repairing the wax combs where honey and pollen are stored and the larvae are reared – and to making honey from the pollen and nectar brought in by older workers. At three weeks, the landmarks of the area around the hive having been learned, the worker embarks on the final stage of her career, foraging for nectar, pollen, resin and water, the basic needs of the hive. She navigates by reference to the sun's angle in conjunction with a built-in time sense, and communicates the place and nature of her finds to the other bees partly by taste and partly by an extraordinary dance, whose movements and figures convey the information by visible means. Hard work finally wears the insect out and she dies.

Are ants intelligent?

Not all bees are sociable, but all their relatives in the ant world are. Ants seem to be even more organised and hardworking, in a way which often alarms as well as amazes the human observer. There are castes of queens, winged males and sterile female workers just as in bees,

The language of bees
A forager honeybee which has found a source of nectar communicates all the details of her find to the other members of the hive by means of a remarkable dance, performed on the comb. If her find is close to the hive – less than 100 yards away – she performs the round dance (1), *her vigour in executing it depending on the richness of the source. The other bees gather round her as she dances, and learn by food exchange* (2) *the scent of the flower and the nature of the find. The dancer is arrowed in this photograph. If the find is further from the hive, she performs the figure-of-eight dance* (4). *On the straight run of this dance she flicks her abdomen, and the number of flicks corresponds to the distance from the hive. The direction of the find is also shown by this straight run: if the nectar can be found by flying directly towards the sun from the hive, then the straight run is performed vertically on the comb* (5), *but if the workers must fly at an angle to the sun, the dance is performed at the same angle to the vertical* (8). *Such is the remarkable organisation within the hive, unsuspected by the casual observer. Other tasks as well as the search for nectar are carried out by the tireless workers. They tend the queen and by licking her* (3) *pick up substances called pheromones which play a part in determining their behaviour. Some workers also look after the growing larvae, feeding them by regurgitation* (6), *and in the autumn they have the ruthless duty of dragging the now useless drones from the hive* (7).

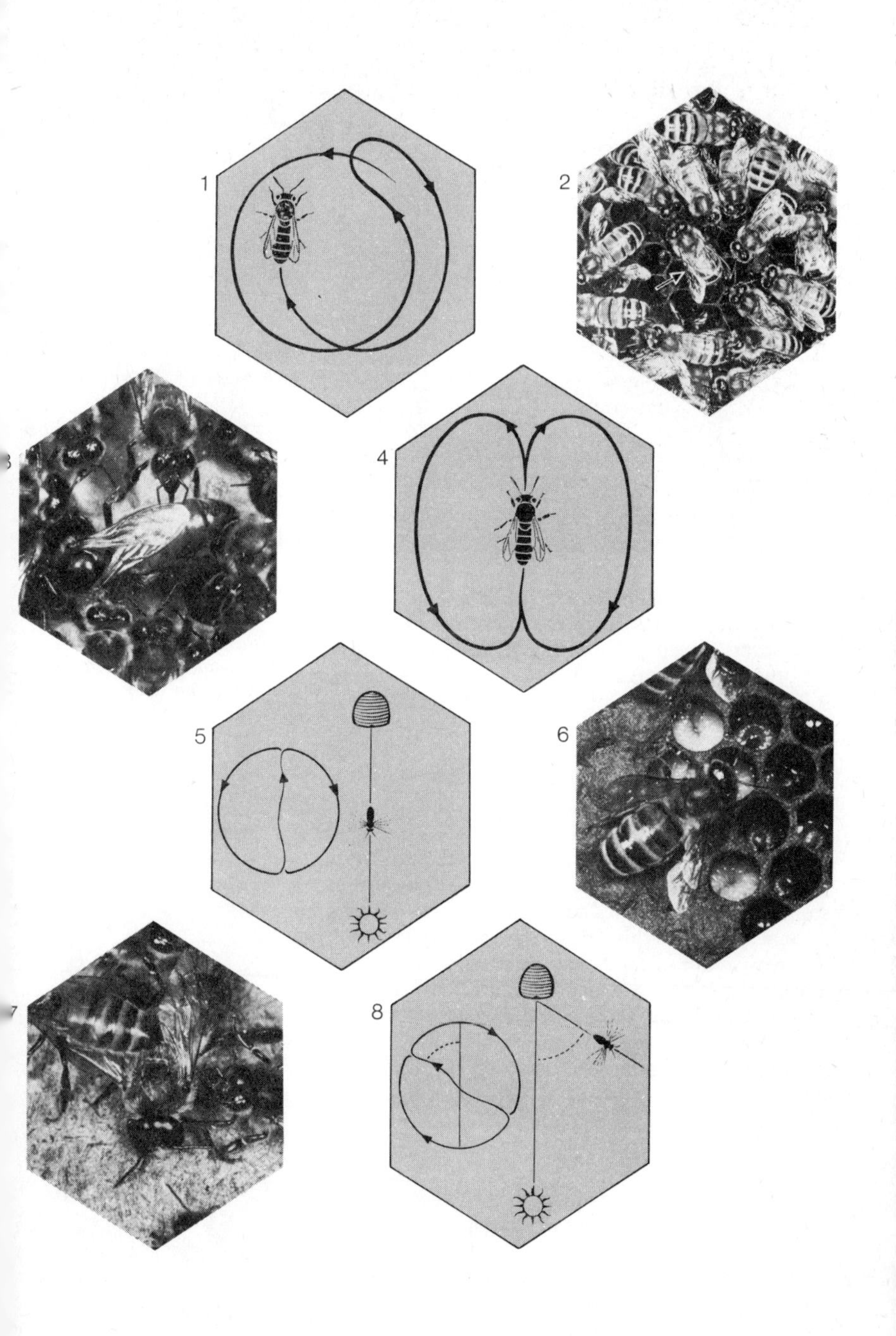
1
2
3
4
5
6
7
8

Farmer ants: leafcutter ants of South America come in three sizes. The largest, called maximae, defend the nest, while the medium-sized mediae collect leaves for the formation of compost. The tiniest of all, the minimae, ride on the pieces of leaf carried by the mediae, in order to ward off attacks by enemy flies.

but some of the workers are specialised fighters, called soldiers. In termites there are soldiers and workers of both sexes, as well as kings, to add to the complexity. On the other hand, they do not pass through a helpless larval stage, and so the task of caring for the young does not arise.

Ants sometimes seem to be enormously clever and are credited with more intelligence than they really possess. This is because there are unseen explanations of their social activities which are still only partially known. When they learn their way through a maze they are certainly using memory to some extent, but they are also laying an invisible scent trail which they and others can later follow. In normal conditions this helps fellow ants to find their way to a new source of food. The apparently tactical flanking movements of a gang of marauding army ants approaching prey are probably an accidental consequence of this. Carnivorous and nomadic, the ruthless army ants seem particularly well organised. They have no permanent nest but actually make bivouacs of their own bodies, hooking on to each other by the feet and making a living system of tunnels and chambers.

Many other ants are farmers and pastoralists; like man they cultivate and store food and even keep livestock. In huge underground colonies of as many as half a million inhabitants, the leafcutter ants of the New World make compost beds from chewed-up leaves collected by the foraging parties. Here they grow a fungus which seems to have evolved in special association with the ants and provides them with their sole food. However, there is no

Left. *All the beautiful coral shapes to be seen in tropical seas are formed by colonies of animals related to sea anemones, together with the hard skeleton secreted by them.*

foresight in their agricultural activities; they are bound by inherited patterns of behaviour and, though an ant's actions within this framework of instinct may appear intelligent, it is quite unable to cope with new situations which require reasoning. Certain veteran members of an ant community may have a 'memory' of former activities and initiate courses of action, but much is done by chemical communication. Substances in the ant's crop, usually derived from licking the eggs and larvae, are passed from one insect to another in the process called trophallaxis or food-exchange, and seem to dictate behaviour appropriate to the colony's needs, so co-ordinating actions in a remarkable way.

Below. *Millepores or stinging corals consist of colonial polyps of two kinds (shown magnified). As in ants, there is division of labour. The feeding type of polyp has a mouth for the intake of food, while the branched polyps which surround the feeding polyps have no mouths but are armed with stinging cells.*

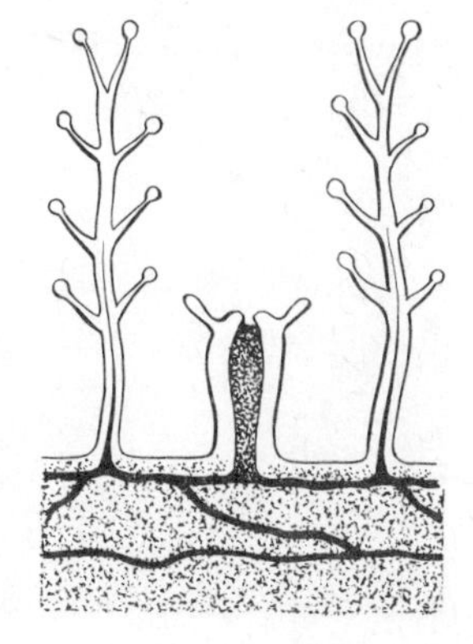

The closest co-operation of all

Even in their closely-regulated society, ants are still clearly separate individuals, but some lower animals are permanently joined in colonies which act as, and look like, single units. *Hydra* is a primitive invertebrate animal, related to sea anemones and plant-like in appearance, which reproduces by budding off new individuals or polyps from the body. If the polyps were to remain attached to the parent, the result would be a branching colony, and this is in effect what happens in the sea firs such as *Obelia*. Corals are more elaborate structures of colonial animals, based on the same plan. Each colony is a continuous system of living tissue or, if there is a chalky skeleton supporting the fleshy part, of interconnecting tubes. All nourishment is thus shared out equally, and as long as a certain number of polyps are obtaining food – usually consisting of small animals – then the rest will not go without, and they are free for other tasks.

Division of labour is sometimes marked, so that there are some polyps adapted for fixing the colony to rocks or the seabed, others functioning in reproduction, and others catching food, often by means of stinging tentacles. The difference in form they exhibit is called polymorphism. Mouthless polyps in the colonies of millepores or elkhorn corals catch and sting the prey and then pass it on to other polyps with mouths to be digested. There may be thousands of polyps in a large coral, helping to create the graceful seaweed-like fronds and other beautiful shapes which abound in warm seas. Clearly the co-operative way of life is a very successful one as far as simple-bodied animals are concerned.

Corals and sea firs, ants and bees, have special kinds of society unique to lower animals. Social life as we understand it is largely confined to the more intelligent animals – mammals and birds – and serves the purpose not just of defence and food-getting, but of providing a secure environment for learning and development in the young of those animals with more extensive capabilities. Obvious practical benefits like these need not be the only reasons for group life. It is easy to forget, when analysing social groups in this way, that some animals stay together for less clearly-defined reasons – perhaps simply because they enjoy each other's company.

Gorgonian or fan corals seem to bloom like flowers by day, as the polyps expand and begin to feed.

Rewards of co-operation

Ostrich stands on stilted thighs –
Panoramic vision from high eyes.
The zebra sniffs savannah air –
Poor lion is thwarted by this pair!

from 'Savannah Animals' by Christabel Epton

The tolerance and co-operation that exists between many animals is something man may envy. Ostrich, antelope and zebra will graze peacefully together. Hermit crab, ragworm and sea anemone will live together in harmony while honeyguide and honey badger form an alliance to go honey-hunting.

Peaceful co-existence is one of the dreams of mankind. But while human beings strive for co-operation and understanding between rival cultures, animal species have long since discovered how to form partnerships and alliances for mutual benefit.

In their natural habitat animals are extraordinarily tolerant towards other species. More and more examples are continually being brought to light of animals living in association with other types, if not actually co-operating in some way. The most stable partnerships are based on a 'fair exchange' system. One animal may be able to provide food while the other offers shelter and protection. Of course, there is no deliberate exchange of services – no such conscious sense of fairness exists in nature. The partnerships evolved from accidental encounters which became regular because the mutual benefits proved to be of advantage to the evolving species. The necessary behaviour has often become 'built in' to the species' genetic make-up, so that the forming of an association is instinctive. Some so-called partnerships are really one-sided, only one of the partners benefiting and the other being indifferent to its presence.

The most complex associations probably began as loose relationships and simple toleration of one another's presence, such as frequently occurs between members of different species today. The animals concerned do not depend on their associates for help and merely take advantage of the aid another type of animal can offer, often without the knowledge of the donor. An example is provided by a tiny African antelope called the dik-dik, which has sharp hearing and a distinctive warning call. When danger – a hunter for example – is present its alarm call is heard by all animals in the vicinity, and they have learnt to recognise its meaning and act on it. In a similar way, animals which herd together by chance take advantage of the particular attributes of their companions. Often ostrich, antelope and zebra will graze together; the long-necked, keen-sighted ostrich can see predators at a distance but has a poor sense of smell, whereas the zebra and antelope have acute noses and can scent a lion hidden in the bush. Baboons and gazelles also recognise each other's warnings, in this case it is the baboons which have good sight and this is coupled with the gazelles' excellent hearing. Each animal knows the other's danger call or signal and they thus have a mutually effective advance warning system.

Animals which share the same environment must tolerate each other to be able to carry out the normal functions of eating, sleeping and reproducing, and often they must live in close proximity when suitable dwellings or feeding areas are sparsely distributed. In a desert, for instance, cacti provide homes for many different species. Birds such as pygmy owls nest in holes in the cactus plants, and rodents

Partners for protection

Baboons and impala form a mutually beneficial alliance in an African national park. The baboons are able to see danger a long way off, while the impala can hear and smell an enemy, so each is able to warn the other.

›urrow in the sheltered areas beneath their prickly stems, vhilst desert insects live amongst the cracks and spines. \ remarkable home is provided by the baobab tree of frica which grows in the savannah lands. Its enormous wollen trunk, designed for water storage, can accommo-ate scores of animals, and a wide variety nest amidst its ranches and in holes in the soft wood of the interior. mong the birds are kingfishers, hornbills and drongos nd they live alongside eagles and owls, which would ormally prey on them. Other inhabitants of the same tree re stick insects, moths, bats and bushbabies. Burrows re also shared in many parts of the world. The prairie ogs of North America build complicated systems, and ther animals – such as burrowing owls, snakes and black-ooted ferrets – take advantage of the tunnels.

Large mixed herds of African game live together peace-bly without competition over the limited grazing avail-ble. They manage to do this because the different species ave developed different food preferences. Gnu like to eat nly young grasses, whilst zebra eat the same grass when it

Who benefits?

Above. *The fluffy chick of a shearwater seems quite unmoved by the presence in its burrow of a venomous Australian tiger snake. The snake gains warmth and shelter, but the advantage to the bird is not clear.*

Left. *Although the white stork is known to eat the occasional small bird, there is no danger to the sparrow which nests in the stork's basement. In fact there is protection in the larger bird's presence.*

Right. *Hermit crab and attendant anemones form the most famous partnershi[p] of all, a true symbiosis.*

is mature and topi antelope feed on it when it is old. Waterbuck, kob and lechwe also practise this form of grazing, so that these three large and closely-related types of antelope can live together in the same area without being in direct competition with each other.

When in the vicinity of their own nests many dangerous predatory animals seem to lose the hunting urge. This has brought about some apparently one-sided partnerships, for some small defenceless species have learned to take advantage of this. They nest nearby, free from attack by other animals, which dare not approach the larger or more dangerous predators. In this way the scarlet sunbird builds its nest in the same tree as hornets, and other small birds often nest near eagle and shrike nests.

The crab and the anemone

Relationships in which different species associate to the benefit of both the partners are called symbiosis. The most involved and close-knit of these relationships occur between the lower animals and mainly in the sea. This is probably because the sea has been a stable environment for a very long time and the animals belong to ancient lineages. Intricate partnerships evolve only over many generations. Sometimes bodily changes take place to further the efficiency of the partnerships and in some relationships one or both of the partners or symbionts becomes dependent on the other for survival.

One of the best-known examples of symbiosis is between some of the soft-bodied hermit crabs, which always live in an empty seashell, and sea anemones. The relationship has been observed in many parts of the world. At first it was believed to be a chance association but it is now known that hermit crabs of several species will regularly make every endeavour to find an anemone and plant it, or encourage it to move, onto the shell. When young, both crab and anemone are free-living, but adult crabs seem to have a strong need to form an association with a specific type of anemone. The anemone accepts the partnership readily. In fact, in one species the anemone will move itself onto the crab's back as soon as it is

Hermit crabs often share their homes with creatures other than sea anemones. Sometimes the lodger actually lives inside the shell, like the ragworm Nereis, *seen here emerging from the back of the shell in search of scraps of food dropped by the crab's mouthparts.*

Opposite; top. *Clownfish sport fearlessly among the tentacles of a sea anemone.* Bottom. *The bitterling has a mutual-help relationship with the freshwater mussel.*
Overleaf. *In their desire to reap the benefits from a plundered bees' nest, the honey badger and the bird known as the honeyguide co-operate in a unique way.*

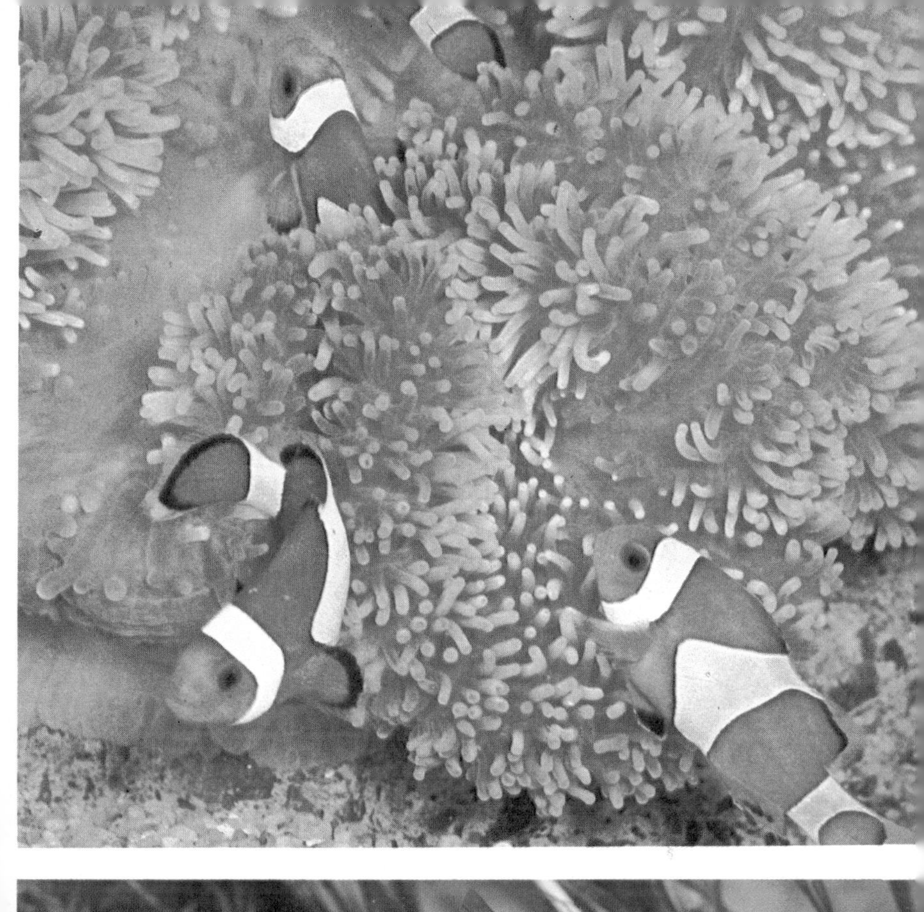

approached. This is quite a remarkable achievement when we consider what a primitive creature the anemone is. It has only a limited nervous system with a few sense cells, and it cannot see in the normal way. But somehow it is able to recognise the right sort of hermit crab and to creep forward, pulling itself up onto the crab's shell and producing a cement-like substance to anchor itself firmly to its new domain. Once an association with a crab is formed the anemone will not leave, and if the shell is outgrown the anemone will move to a larger shell with the hermit crab.

Particular kinds of anemone associate with particular species of crab. The various species have differing degrees of independence. The cloak anemone *Adamsia palliata*, found around British coasts, cannot survive if it is removed from the shell of its associate Prideaux's hermit crab. This crab chooses only a very small shell, so that much of its body is unprotected. It seems that the shell serves merely as an attachment for the cloak anemone, whose basal disc extends to cover the crab's body. One benefit is that

In spite of the apparent danger from the stinging tentacles, this little fish lives unscathed in the shelter of the large man o'war jellyfish.

the crab does not need to expose itself during development by changing shells like other hermit crabs, as the cloak anemone grows at much the same rate and keeps it well covered. The association between these two sea creatures is evidently highly developed, but it is not really clear what the hermit crabs and anemones gain from it. Certainly the anemone provides the crab with a degree of camouflage and protection, but this is not always effective as some species of fish have been observed to eat both crab and anemone. It has been suggested that they feed together, the crab permitting the anemone to take scraps of its meals, but generally they eat different types of food and often the anemone's mouth faces away from the crab's. However, the anemone gains by being transported to new feeding grounds by the crab, which is definitely an asset as on its own the anemone is more or less stationary.

Hermit crabs are often camouflaged by other sea animals with which they form almost accidental associations, such as sponges, barnacles and sea firs. They also associate with a ragworm *Nereis fucata* which conceals itself in the terminal whorls of the shell, coming out when the crab feeds and snatching pieces of food from the crab's mouthparts. It would be easy for the crab to eat this worm but for some reason it tolerates the rather one-sided arrangement and allows the worm to remain in the shell unmolested.

The arrow goby of California shares holes in the mud flats with a pea crab. The goby acts as the hunter, returning to the safety of the hole with large items of food that the pea crab tears up for consumption; the goby gets the scraps and tiny particles left go to the burrowing worm whose hole they all share – co-operation in triplicate.

Death-defying fish

The larger anemones have further relationships with other sea creatures. One of the most common and also most interesting is the association of some very large species with the brightly-coloured little clownfish, *Amphiprion*, which is found around tropical coral reefs. To avoid predators, most species of this fish take up their stations near crevices in rock or coral into which they can dart when alarmed. Some clownfish, however, use anemones as their protectors, despite the fact that a large

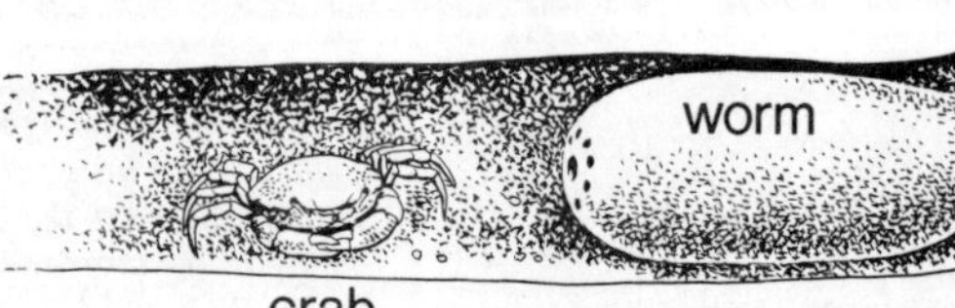

anemone normally eats fish. The clownfish swims among the anemone's tentacles, even sleeping among them in apparent safety. It is not attacked because it apparently 'makes friends' with its particular host and is then able to live happily in its protection, even producing a brood of young in the shelter of the dangerous stinging tentacles.

Experiments have shown that the clownfish is especially adapted to form this association. It is covered with a thick, slimy mucus which prevents the anemone's sting from being triggered off when the clownfish makes its first overtures of 'friendship'. After the clownfish has circled the anemone several times it is accepted and the anemone makes no further attempts to attack it. In technical terms, it habituates the anemone to its presence. The relationship is now established. It has been suggested that in return for protection the clownfish lures food within reach of the anemone's tentacles, though this is probably accidental. Other ways in which it is said to help are by aerating the water or removing irritating waste food from inside the host: the fish do indeed disappear into the anemones 'stomach' and come out again without being harmed.

This type of association also occurs between species of small fish and the large jellyfish. The Portuguese man o' war shelters a small fish called *Nomeus gronovii* amid its long stinging tentacles. The fish occasionally robs the jellyfish of food, and although it is not totally immune

Above. *Keeping company with sharks is the unusual behaviour of pilot fish. They were so named because they seemed to direct the course of sharks and other large fish, but there is probably no foundation for this belief.*

Opposite. *In mudflats in California another curious alliance – between a marine burrowing worm and its lodgers – may sometimes be observed. The worm feeds by secreting a slime net and drawing a current of water into its burrow which brings both food particles and oxygen with it. The food is trapped in the net. Deriving benefit from the shelter of the well-aerated burrow may be several other animals, like the arrow goby and pea crab shown here, as well as other worms and clams.*

So well adapted is the remora to hitching rides on other fish that its dorsal fin has become modified to form an attachment sucker.

to the poison from the jellyfish, it can withstand more than ten times the dose lethal to another fish the same size. Why fishes like *Nomeus gronovii* choose to shelter in such a dangerous place is not known, but the habit probably originated from hiding beneath non-living, floating objects such as driftwood.

An extreme example of association for protection is seen in the behaviour of a small crab *Hapalocarcinus marsupialis* which lives on the Great Barrier reef. Many small animals associate with coral or sponge for the shelter they afford. The female of this crab has taken this a stage further. She takes up residence in a fork of coral and sets up a flow of water which causes the coral to grow around her, so that she is eventually encased in a marble-sized cage and is completely safe from predators. Water can enter through tiny holes in the coral and, although the female is totally imprisoned, males of the species can visit her as they are very much smaller.

Undersea cleaning service

Many partnerships exist between fish which seem to need company. Pilot fish doggedly swim with sharks and other large species but, although they may well benefit from living in the shark's vicinity by gaining food scraps and protection, they will transfer their allegiance to a ship or other large object if the shark is killed.

Lethargic fish will actually hitch-hike by riding or attaching themselves to the bodies of bigger fish, thus gaining fast transport to new feeding grounds. In some cases these relationships are mutually beneficial as the hitch-hiker will eat parasites on the host's body. One fish which is specially adapted for hitch-hiking is the remora, which attaches itself to its host with a powerful sucker, actually a modified dorsal fin. It is interesting to observe that not only has this fin evolved as a sucker but the remora's sensory organs have also been adapted to synchronise its movements with its larger partner. Larger species of remora probably leave their hosts to feed on passing shoals, but small types have been noted inside the mouth and gill cavities of large manta rays or sailfish, and it is likely they help the host by feeding on parasites.

The fish world has many associations and scientists are only now finding out the significance of these relationships. One of the most important is the relationship of many fish with cleaner or barber fish. These fish provide a 'doctoring' service by eating the parasites, patches of dead skin, bacteria, fungus and fish lice which infest their hosts. In return for this service the cleaner fish get plenty of food, and because other fish welcome their attentions they are not attacked, even by large, dangerous fish.

Distinctively striped, the barber fish Labroides *is well known for its cleaning activities. The customer here is a surgeonfish.*

The importance of the services of cleaner fish has been demonstrated in marine aquaria. If cleaners are removed, the other fish become seriously infested and on the return of the cleaners they will queue up for treatment. This is a reflection of what has been proved to happen in the sea; there are actual cleaning stations, usually marked by a bright piece of coral, where the cleaner positions itself. Fish from deeper water come to this spot for treatment, and they hold themselves rigid, with fins extended, while a cleaner goes over the body, entering the gill chambers and even the mouth. Some fish are cleaners only when young, but the true cleaners give lifelong service. They advertise themselves by their brightly-patterned bodies and have special long, pointed snouts and tweezer-like teeth for digging out parasites and fungus. Cleaners in temperate seas are less brightly coloured and often follow their clients rather than having a regular base; they may work in teams to clean a big fish such as a shark.

There are over twenty different species which act as cleaners, including wrasse, surgeonfish, angelfish and a goby as well as six species of shrimp and a crab. One of these fish, the cardinal, cleans only a sea urchin; it removes waste material from between the urchin's spines. At the slightest disturbance the urchin spreads its spines defensively and the fish shelters head down amongst them.

Helpers and hitch-hikers

These fascinating associations are not confined to the sea. In some European and Asian rivers a close relationship has evolved between the freshwater mussel and a fish, the bitterling. The female bitterling spawns in spring, laying her eggs inside the siphon of a freshwater mussel. Water being sucked through this tube draws the eggs into the

Carmine bee-eaters have long been known to perch on the backs of kori bustards, which act as beaters, but recent reports describe how they are transferring their allegiance to motor cars, which serve the same purpose.

mussel's gill cavity. There they develop, benefiting from the protection of the mussel's body and being constantly aerated by the water flowing through the siphon. At the same time as the fish spawns, the mussel discharges its larvae, which attach themselves to the fins of the adult bitterling. Once they take hold they are encysted by the bitterling's skin, remaining there and feeding on the fish's body fluids for three months until fully grown. They then fall off, to start new colonies of mussels and spread the species to new parts of the river. The mussel, however, although greatly aided by having the bitterling on hand, does not rely on it to spread the larvae. The amazing part of the relationship is that it is not only the presence of the male bitterling but also that of the mussel which stimulates the female to spawn. Without the mussel the female bitterling has no urge to lay, and thus the species is totally dependent on mussels for its continuance – an extraordinary dependence which is difficult to explain.

Generally speaking, land animals tend to have more casual relationships, the individuals involved gaining from the co-operation of different species without being totally reliant on it. Some partnerships are similar to those in the sea. There are many examples of hitch-hiking by smaller animals which have very limited powers of locomotion. They do not harm their host and are usually tolerated with what amounts to animal good nature, as the host rarely receives any benefit. Mites and false scorpions, for example, frequently help themselves to lifts from harvesters, bluebottles and daddy longlegs. Woodlice have been known to benefit occasionally from lifts on flying beetles and certain mites gain access to the ants' nests where they scavenge by travelling as passengers on the legs of the swarming ants.

A hitch-hiking relationship is found between the large kori bustard, an East African bird which may grow to over five feet in height, and the small carmine bee-eater. Picking its way through the grass, the bustard disturbs many insects, and the bee-eater merely needs to make short trips from its perch on the bustard's back to reach plentiful food. This is a rather more developed version of the relationship formed by cattle egrets and other birds with grazing animals. These birds also snap up insects disturbed by their hosts, but they do not ride quite so regularly on their backs.

Birds which eat insects are commonly found to associate with large animals that are likely to carry parasites. The European starling associates with sheep, sitting on the wool and removing ticks from the sheep's back; aquatic birds associate with hippopotamus, using the great beast as a resting place and fishing platform, and pecking ticks from the hide. A small bird which normally feeds on plankton, the grey phalarope, will land on the backs of whales that surface nearby and make a meal of the whale lice, small crustaceans which infest sperm whales. Marine iguanas have a similar loose association with rock crabs. The crabs scuttle over the iguanas' bodies as they sit in the sun and pluck ticks and the occasional piece of skin from the reptiles, ridding them of parasites.

One quite close relationship is found between the Egyptian plover and the crocodile. The plover takes ticks, leeches and scraps of food from the crocodile's scales and even from inside its mouth, as the reptile lies sunning itself with jaws agape. Besides cleaning its skin the plover warns the crocodile of danger by flying away, and if the crocodile is not alerted by this it will return and peck it sharply on the head.

Two African specialists

So far we have learned of the associations between land animals which are well established but not very intricate. No physical adaptations have been necessary. It is easy to see how an animal will quickly acquire the habit of a partnership which means little alteration from normal behaviour but makes life for the animals concerned much easier.

However, there are terrestrial animals that have become physically adapted or have evolved special patterns of behaviour to enable them to associate with another animal in a more complex relationship. A bird which is a fine example of this is the oxpecker, which has a strong relationship with game – especially larger African animals such as buffalo and rhinoceros. The oxpecker is believed to form an association with one particular animal in a herd, not necessarily choosing an animal heavily plagued by ticks (upon which it feeds). It stays constantly with this one animal, conducting most of its life – feeding, courtship and mating – on its back. The oxpecker is well adapted to this life. It has sharp, curved claws for gripping the hide of

Cattle egrets follow cattle and game around in search of insects disturbed by the larger animals' feet, and they also take advantage of the perches offered by the sturdy backs of animals such as rhinoceros.

its mobile partner and a long stiff tail which acts as a support when the bird is perched on the legs or flanks. Its beak is flattened and short and is often used with a scissoring motion whilst laid alongside the hide, so that it can efficiently tweak out parasites, though it may at the same time enlarge wounds to feed on the blood. Like the Egyptian plover, the oxpecker serves its host by making repeated warning when danger is present, calling and flying low over its head.

Another African bird, the greater honeyguide, has learned to co-operate with the honey badger or ratel in a remarkable relationship, although neither possesses any special modification of the body to facilitate the partnership. The honeyguide feeds on beeswax, an unusual food indigestible to most animals, while the honey badger, as its name suggests, is a honey eater. Somehow the honeyguide learned that the badger's ability to rip open bees' nests with its powerful claws could be put to use. When the small bird sees a honey badger, it attracts attention by flying up and calling. Leading the badger on in this way, the honeyguide eventually arrives at a tree containing a bees' nest. Once there the bird stops calling and the badger locates the nest, tears it open, and exposes the wax cells filled with honey which make up the honeycomb. Then, after the badger has eaten all the honey it wants, the honeyguide is left in peace to peck away at the beeswax so conveniently exposed.

Insect opportunists

The social insects – bees, wasps, and especially ants and termites – form hundreds of different associations with other insect species and sometimes with larger animals. Their highly-organised, well-regulated societies allow plenty of opportunity for the development of partnerships, 'hangers-on' and slave relationships. Just as civilised human society attracts animals such as rats, sparrows and pigeons to take advantage of the food supply and shelter provided by buildings and towns, so the big insect societies harbour their own share of opportunists. Some of these can be put to use in the same way as domestic animals are used by man, while others are merely tolerated.

Bees and wasps are normally very intolerant of intruders, but they have come to accept certain hoverfly larvae in their nests. These larvae feed on the debris of the colony, so performing a cleaning service and receiving in return the benefit of food, shelter and the protection afforded by their dangerous hosts.

Some African and Australian species of termite build huge nests of mud called termitaria, which are extremely solid, waterproof and well ventilated; ants also build comfortable nests for their colonies, and both types of structure may contain 'guests' of many sorts. Frequently these are other insects which act as scavengers, collecting and consuming waste materials from the nest. Others form a rather one-sided partnership, of benefit only to themselves. Wingless insects called silverfish, which often live in ants' nests, take advantage of their hosts' habit of passing food from mouth to mouth by snatching a morsel for themselves as the exchange takes place.

Termites also pass food in this way. The tiny mahout beetle, a relative of the devil's coach-horse beetle, rides on the head of a termite, reaching forward to take its share of the food. This extraordinary behaviour earns the beetle its name; a mahout is the boy who rides a working elephant in India. Whether the termite gets anything in return for the beetle's thieving is not certain, but it might be that the mahout is one of the many insects tolerated by ants and termites because of the sticky substances they secrete. These are the basis of a whole range of partnerships, of which the most famous is that of certain ants with aphides, the greenfly and blackfly so detested by gardeners.

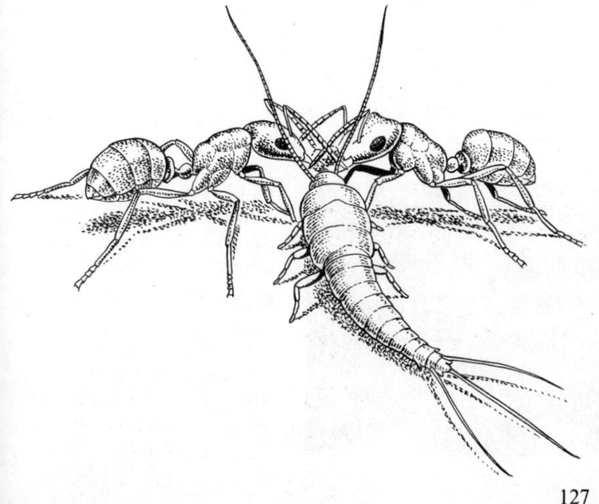

When two ants exchange food, they are not always quick enough to prevent the scavenging silverfish from taking its share.

Ants as farmers

This red-and-yellow ground barbet of East Africa has broken the defences of a large termite mound and now nests there, sheltered and protected.

The aphides feed on large quantities of sugary sap from plants, and as waste they secrete a sweet, sticky substance called honeydew, to which ants are extremely partial. Thus wherever aphides congregate on plants ants are found greedily licking up the secretions. However, this is not all; because ants crave this substance they tend the aphides, carrying the aphid eggs into the ants' nest and caring for the young. They will also 'herd' the aphides onto the best sources of food, even carrying them to the chosen pasture. By taking the honeydew as it is produced they prevent this sticky substance from killing the plant and even sometimes from drowning the aphides. The ants 'milk' the aphides by stroking them with the antennae; this encourages the aphides to secrete the honeydew.

This very involved relationship helps both sides. The aphides develop and multiply more rapidly and are protected from the insects that harm them, such as ladybird larvae. In return the ants receive increased quantities of the honeydew they love. In fact, the association has developed so far that some species of ant live on honeydew alone and do not eat any other food material at all.

Ants also associate with a honey caterpillar and a rove beetle to obtain their secretions. These two partnerships, however, are rather different from the 'farming' of aphides. The ant becomes 'addicted' to their secretions, which are not food and are taken rather in the way that people drink alcohol. The ant community as a whole suffers from this partnership, as the individual ants bring the caterpillar and rove beetle into the colony, where they are pampered and fed and even permitted to feed on ant larvae. Sometimes a whole colony is wiped out as a result of the honey caterpillar being introduced into the nest – definitely too much of a good thing.

Larger animals also associate with ants and termites. Because the termitaria are so strongly constructed, many animals find them excellent shelter. Kingfishers, parrots and other birds manage to excavate nest holes within the tough walls, and the termites merely seal off the damaged parts of the nest, so that their carefully ventilated and insulated home is not affected. Reptiles as well as birds

Left. *A droplet of honeydew exudes from the abdomen of an aphis as it is milked by a wood ant 'herdsman'.*

Right. *Ahla the baboon made the headlines in her native Southwest Africa when she took over the duties of goat herd. Here she is grooming a young kid from her flock.*

have been found in termitaria. The Nile monitor, a large lizard, has discovered that the warm, well-aired conditions inside are ideal for incubation. The female tears a hole in the termitarium wall when it is softened by rain, lays its eggs inside and leaves them to develop. The termites repair the damaged spot, completely sealing in the eggs so that they are quite safe until hatching time. Fluid from the eggs is believed to soften the walls so that the young are able to scrape their way out.

Strange friendships

Occasionally we learn of odd partnerships between animals which are normally enemies or, at least, which do not form partnerships in their wild state. These relationships occur most often in zoos where the animals may be bored and lonely, and in some cases are due to the frustrated maternal instinct discussed in an earlier chapter. In a zoo especially, a young animal may be 'imprinted' on a completely different species, coming to believe itself a member of the species with which it is caged. A white peacock reared with Galapagos tortoises in a Vienna zoo would take no notice of peahens because it had spent all its life with the tortoises and firmly believed it was one. There is a report from an African farm of a hand-reared female baboon which was kept with a goat flock. The baboon adopted one of the goats as mother and rode on her back at first, but as she grew older she began to take over the leadership of the herd. She looked after the goats with remarkable devotion, making sure the young were fed, frightening off any strangers who approached too closely and generally performing the functions of a good goat-herd.

Man can train animals which do not usually associate to live peacefully together. Dogs and cats are the most common example; a dog will have a companionable relationship with a cat in the same house but still chase other cats in the neighbourhood.

Other animals are often put together because one has a calming influence. A chicken put with an elephant during air transport will keep it happy and peaceful. Donkeys, very stolid, calm animals, are put in the same field as excitable racehorses, and a billy goat kept with cattle helps to keep them placid.

In nature such occasional associations between individuals are very rare. Animals will form associations only when there is something to be gained from it by one or the other partner, and if one member of a species enters a partnership, the likelihood is that they all do. In time the association becomes an integral feature of the behaviour of a particular type of animal, as much a part of its way of life as the way it feeds, moves and reproduces.

Man tames the wild

In time the savage bull sustains the yoke,
In time all haggard hawks will stoop to lure.

from 'The Spanish Tragedy' by Thomas Kyd

Human life would not be complete without animals. Perhaps the most important are the domestic species which serve us. Many people derive much pleasure from keeping pets or taming wild animals, while others seek to gain understanding of wild creatures by living with them in their natural habitat.

In Tanzania a young Englishwoman studies the ways of wild chimpanzees by living with them in the forest. On the same continent, the vast herds of game animals are surveyed for possible use as a new source of human food. Across the world, a North American couple spend nine years filled with joy and heartbreak sharing their home and their life with a group of captive wolves.

These are some of the ways in which people involve themselves with animals. Man is continuously coming into contact with animals, from the game he hunts to the pets he keeps, the horses he rides and the livestock whose flesh he eats. Since he first appeared on the earth, he has tried to dominate, tame and sometimes eliminate the animals which share his environment. Now, as he realises the devastation caused by some of his activities, he is increasingly concerned with trying to restore the balance between making use of animals for his own needs and preserving what is left of wild nature.

Instead of imposing human wishes and standards on animals, people have sometimes tried to enter the animal's world in a peaceful quest for understanding. They follow the tradition of the early thirteenth-century saint, Francis of Assisi. A friend to all creatures, he is said to have had a great love for animals, and preached both about them and to them. People who have the ability to attract wild animals and be accepted by them certainly seem to exist. There are those who claim that wild birds will fly down and perch on their shoulder, or that deer and other shy creatures show no fear in their presence. The shyness of animals can sometimes be overcome by talking softly and avoiding jerky movements. A direct gaze is also distrusted by animals, particularly birds, to whom staring eyes represent a definite threat. But for a real understanding, the observer must shed his human way of thinking and try to put himself in the animal's place, learning and recognising the signals and sounds by which members of a species communicate. This is what the pioneer behaviourist Konrad Lorenz did, notably in the case of a colony of jackdaws, started by himself but allowed to live free. His amusing and enlightening experiences are told in the book *King Solomon's Ring*, probably the liveliest introduction to animal behaviour ever written.

More recent achievements have concerned scientists anxious to learn more of the behaviour of species in the wild. Some remarkable studies on large mammals have been produced by researchers who deliberately tried to enter into the everyday life of the animals and gain their confidence. Living on the shores of Lake Tanganyika, far from any human settlement, the young English zoologist Jane Goodall carried out a study of several years' duration on the chimpanzees of the Gombe Stream Reserve. She learned to recognise all the important members of the local group of chimps, and gave them individual names.

Keeping a supply of bananas on hand proved a good way of making friends, and eventually the animals became used to her quiet presence, so that she was able to observe and record their habits, development and relationships with each other. In one moment of triumph, a wild chimpanzee mother showed her trust for Jane by bringing her new baby with her into the camp. In another, an adult male named David Greybeard allowed her to perform the friendly gesture of grooming him. Along with the younger animals in the group, this affectionate chimp would even play with his human friend.

Following Jane's example, an American called Dian Fossey left her job in California to spend three years with wild mountain gorillas in central Africa. She overcame their shyness by imitating gorilla noises and actions, including chest-beating, which may have looked a little undignified but seemed to convince the gorillas that she was one of them. The result of her patient observation was a valuable study of the habits of one of man's closest relatives.

Apes in the house

Many people enjoy the presence of a tame animal in the home. Their reasons for keeping a pet may range from genuine affection for animals to a desire to rule another

Prof. Konrad Lorenz, the scientist whose pioneer work, particularly on geese, was the foundation of the modern study of animal behaviour.

creature or mother it like a child. Observing the life of an animal at close range is also a valuable educational experience for children. Much can be learnt in this way about the development, reproduction and behaviour of animals. Of course, zoos have the greatest use in this field and they also do valuable work in research and conservation, but we shall be concerned here with the less well-known activities of some private individuals.

Pet-keeping is not confined to civilised societies, and seems to be a long-established aspect of human nature. Many primitive tribes care for orphaned animals and other creatures which seek food and shelter near their homes. Apart from the old favourites like cats, dogs and rabbits – animals bred specially for captivity – there is a growing vogue in many countries for keeping wild animals as pets. Where these are potentially dangerous or liable to be maltreated, there is need for a review of the regulations concerning their care and control. Tame foxes, badgers and even lion cubs have become so commonplace that they excite little attention. The more ambitious owners go in for adult lions, leopards and alligators. Often this is purely for the excitement of owning a dangerous or unpredictable animal, but sometimes there is a more serious side to the venture. John Aspinall, who

Hyaenas may not seem the most attractive of animals, but here is one man, living in Ethiopia, who has made friends with an entire wild pack. He has gained their confidence to the extent of being able to feed them in this unusual way.

Opposite. *Round-up on a cold and frosty morning. The Lapps do not need to feed or stable their reindeer but they gain from them milk, cheese, meat, hides and bone.*
Overleaf. *Easily tamed, kestrels are useful for the novice falconer to practise caring for his birds, but they are not easily persuaded to take quarry.*

numbers gambling, travelling and nature conservation among his many activities and interests, has set up colonies of rare species at his country home in the south of England. The clouded leopard, tiger and lowland gorilla are three of the endangered species kept by him which may be saved from extinction by breeding in captivity. The gorillas of John Aspinall's colony have not yet bred, but have settled down well in the presence of human beings. The smaller animals play and wrestle with their owner, and go for walks with him. The largest, named Gugis, has come to consider himself leader of the group; play with him is no longer advisable, because of his jealousy of human rivals.

Another ape, whose survival as a species is even more in danger than that of the gorilla, is the orang utan of Borneo. This appealing red-haired forest dweller has been all too popular in zoos in the past. In order to capture very young animals for export, unscrupulous dealers anxious to cash in on this popularity caused the mothers to be shot, so that the number of animals in the wild capable of reproducing steadily diminished. Some of the youngsters illegally kept by natives found their way into the care of Tom and Barbara Harrisson, of the Sarawak Museum. Their home in Kuching became a centre for rearing orphaned orangs. Treated as human infants in many ways, the apes were also allowed to play and practise climbing in the trees of the garden. Most had to be sent to zoos when they were older, but the Harrissons cherished the hope of one day returning some of them to their natural home in the forests, where they might strengthen the position of the wild population. At length, two candidates for the experiment, Arthur and Cynthia, were introduced to a large enclosure built for them in the forest. Although they had some success with Arthur, who showed a marked interest in exploring his new surroundings, Cynthia was reluctant to leave the security of her human foster home. Orangs reared in captivity become used to human ways and will always tend to be over-trustful of people. This increases their chances of being shot by hunters after their release into the wild. In addition, they have not had the benefit of an upbringing within an orang group, which teaches them more about survival than they can ever find out for themselves.

Gentle giants

Lions, though they appear ferocious, are among the most easily and successfully tamed of all large wild animals. This is because they are basically very lazy, and always willing to accept the life of ease which captivity offers them. In the wild they are not very active except when driven by hunger to make a kill, and are most contented when

Left. *Boy and Girl, two formerly captive lions, in playful mood with Kenya game warden George Adamson.*

Right. *An excited hug from a grey wolf for Dr. E. H. McCleery of Pennsylvania, who at the age of ninety would still walk unafraid into the wolf pen. He had persuaded the authorities to let him keep the animals caught or injured in an extermination drive, and they eventually accepted him on friendly terms.*

stretched out sleepily in the sun or the shade of a tree. To keep such animals in restricted cages, such as are found in many zoos, is far less cruel than to confine an animal such as the wolf in a cage too small to run about in, because wolves have the urge to travel long distances. On the other hand, no animal should be kept in a small cage when there is space available to give more freedom, and most zoos – especially the open ones – realise this.

Lions live and breed well in captivity, and there are numerous examples of full-grown animals being kept in the home. The most famous of captive lions was, however, kept in much closer contact with her natural habitat. She was Elsa, whose story was described in the book *Born Free*. Game warden George Adamson and his wife Joy took over the rearing of three very young cubs after their mother had been shot in the Kenya bush. Though two of the growing cubs had to be sent to a zoo, the youngest and weakest, Elsa, remained as a playful addition to the household. She was able to wander free in the surrounding bush, but she became used to human ways and very attached to her adoptive parents. When she was almost full-grown, it was decided that she should be freed. Repeated attempts to abandon her in a suitable area away from human dwellings and close to other lions met with no success, and invariably ended with the lioness strolling back into camp with a distinctly hurt expression on her face! Unable to feed herself and shunned by the wild lions, she was simply not ready for such independence. She had to be introduced to wild prey and persuaded to

make her own kills before, at three years of age, she finally began her life in the wild. Although she found a wild mate and had cubs of her own, she always remained tame and would greet her former keepers joyfully when they came to visit her.

Elsa was accustomed to roam the bush from childhood, but her successors Boy and Girl had been born and raised in captivity, and had never seen the open spaces when they arrived at the Adamsons' home at the age of nineteen months. They had been mascots of the Scots Guards before starring in the film of *Born Free*, and George Adamson felt it would be appropriate to attempt their training for a return to the wild. In a bid to awaken their natural instincts and build up their confidence and independence, he took them for long walks and hunting expeditions, until they had learnt the skills necessary for survival. Always tame and friendly towards humans, they nevertheless became able to look after themselves and were accepted by wild lions, a considerable success for their trainer.

A story with a different ending comes from North America, where Cris and Lois Crisler attempted to rear five orphaned wolf cubs which they adopted during a visit to Alaska. While in their natural surroundings, the wolves were allowed a certain amount of freedom, but when the time came for the Crislers' return home it was clear that the cubs were not yet able to fend for themselves. Rather than leave them behind to face starvation, the Crislers took them back to their mountain home in

Colorado. Here a series of large pens were constructed to give the cubs as much freedom and variety as possible, and 'walks' up the mountain were arranged for exercise.

Despite all their efforts only one, a female called Alatna, survived. The others escaped and were later shot. Alatna lived on, showing great affection towards her keepers but obviously pining for the company of other wolves. So that she could start a pack of her own, the Crislers obtained several dogs to be her companions, and she eventually mated and produced several litters of wolf-dog puppies. In this way she fulfilled her natural instincts in spite of her captivity. Lois Crisler describes her successes and failures with the growing colony in her book *Captive Wild*, revealing her increasing understanding of the wolf's social nature and needs during the nine years of Alatna's life.

The beast acknowledges his master

Animals have been tamed both for pleasure and for more practical use. The mastery of lions and tigers in circuses is an end in itself – a demonstration of skill and a source of entertainment. It is not, as are many acts of taming, the means of achieving as an end product an animal or species which will serve man, perhaps as a beast of burden, or a companion, or a food source. An enormous range of species can be and have been tamed, from tigers to whales and including unexpected animals such as ostriches and extremely wild forms such as zebras.

To a certain extent, tameability is inherited. Some forms are naturally tamer than others, and within the species there is also individual variation in this respect. A man training circus animals has to meet the problem of the

Leaping effortlessly over a high wire, a trained dolphin demonstrates its acrobatic skill.

differing characters of animals – some are ferocious, others sly or docile – and he must 'play them at their own game', learning to recognise the indications of their moods. His aim is to make them submit to his will. The usual method of persuading animals to do what the trainer requires is to keep them hungry, so that the promise of a food reward will lure them to further progress, though this is too dangerous a procedure to use on the big cats.

Firstly, the trainer must establish his dominant position by showing his confidence and power and becoming, so to speak, the top-ranking animal in the group. Then he may invoke the animal's sense of territory, by making it understand that a particular part of the cage, such as a stool, is its own domain, where it will be left in peace. When it returns to this spot it is praised and rewarded. Another technique is to play upon the wild animal's habit of allowing a dangerous being to approach to a certain point before fleeing. This can be developed so that a slight movement suggesting approach on the part of the trainer results in the animal moving away without being touched. When cornered in a cage, the animal will turn to attack if provoked. Using his whip, the trainer induces the animal to rear up in threat and then suddenly, when this is done, the provocation ceases and the animal is praised. If this is repeated often enough, the lion or tiger learns to perform this trick at the slightest movement of the whip. The trainer is exploiting a natural response and modifying it so that it becomes fixed and ritualised by repetition. He is aided by the ability of some animals to recognise moods by learning the meaning of movements almost imperceptible to man. In the same way, a dog knows whether his master leaves the room because he means to go for a walk or because of some other motive of no interest to the dog. The performance seen by circus-goers is the culmination of

Falconry was long the sport of kings and noblemen in England. This print shows King James I hawking.

much work based on this kind of animal psychology, and consists of many acts of learning linked together to make a continuous whole.

A new showplace for the animal trainer's skill is the seaquarium, where dolphins and whales are put through their paces performing tricks such as playing ball and leaping through hoops. Part of the immense popularity of seaquaria derives from the apparent enjoyment shown by the animals. They will not perform unless they want to, say the trainers. But animals have been trained with more serious purposes in mind for thousands of years; their natural activities have been curbed and channelled to suit man's needs. Falconry or hawking, the art of using birds of prey in hunting game such as rabbits and grouse, originated in the East over three thousand years ago and is still a major skill in Persia and Arabia. It became very popular with the English and French aristocracy in the Middle Ages – until superseded by the widespread use of firearms – and the archaic language surrounding the practice of falconry reflects this period of its greatest vogue. Strictly, a falcon is one of the long-winged species of open spaces, such as the gyrfalcon, peregrine and merlin, while a hawk is one of the short-winged woodland birds such as the goshawk and sparrowhawk. Young birds may be taken from the nest or eyrie for training, in which case they are called eyasses, or they may be caught when adult, and are then called haggards or passage hawks according to their age. Mastery over a hawk is gained by keeping it hungry so that it will fly to a lure of flesh or to live quarry, but the actual technique in hunting depends on the type of bird and its prey. For ease of handling and during travel, the bird is blindfolded with a hood or cap of leather to keep it calm, and this is removed when the quarry is sighted. Jesses – short leather straps on the legs to which a leash is attached, and bells for locating the bird, complete the basic equipment.

The cheetah has also been used in hunting, serving to run down deer and other game rather like a greyhound. In the wild it is found in much of Africa, ranging eastwards as far as India. Unlike the leopard, it is friendly and easy to tame, as the Egyptians and Assyrians found when they first employed its services three thousand years ago. The practice spread to central Asia and India and is continued to this day. Because of the cheetah's poor endurance, it was carried on horseback behind the rider until the prey was spotted, and then allowed to jump down and set off in pursuit.

Although cheetah cubs were sometimes taken from their mother at an early age and tamed for the chase, the most successful hunters were those caught as adults – after they had learned the skill of the chase in the wild. One Mogul emperor is described as having a thousand such hunting cheetahs in his menagerie in India.

In a famous painting by George Stubbs of a cheetah with two Indians, the tame beast is shown harnessed and ready for the pursuit of deer and other game.

A well-trained tusker at work in a teak forest in northern Thailand. Working elephants are a much more common sight in Asia than they are in Africa.

Top. *The Egyptians tried domesticating some unusual animals. This Old Kingdom scene from a tomb of 2500 B.C. shows oxen* (top) *as well as tethered gazelle, addax, ibex and oryx* (centre) *and hyaenas for fattening* (bottom).

Above. *An unusual experiment took place in South Africa in 1892, when zebras were used to draw stagecoaches.*

Opposite. *A zebroid safari in Kenya. These hybrid beasts of burden are more readily tamed than zebras.*

The taming of the species

Cheetah and falcon cannot truly be described as domestic animals. Although tamed they are not deliberately bred for their task. Domestication can be thought of as the permanent taming of a whole species rather than of an individual. Breeding animals, which will become the parents of the next generation, are selected for traits such as docility and use to man, so that the species is subjected to a sort of man-made evolution which may totally change the animal's appearance and habits. Virtually all the animals which have been domesticated are social species, which means that they are able to form bonds with each other and also – and this is most crucial – with man. Domestication represents a partnership exactly parallel with the symbiosis described in the previous chapter. The cow, sheep, hen or whatever it may be provides man with milk, leather, wool, meat, eggs and so on, and in return is fed and protected in every way. Coinciding with the development of agriculture in the New Stone Age, domestication represented a revolution in the life of man. Now he was assured of a regular supply of food, clothing and beasts of burden. Together with the beginning of crop cultivation, this meant that he could settle down in

one place. From these settlements arose the great civilisations of the world.

Most of today's domestic animals are derived from wild stock domesticated – usually deliberately – between four and nine thousand years ago. The dog, perhaps descended from the wolves which used to hang about the camps of primitive man, seems to have been one of the earliest. Unlike the sheep, goats, pigs and cattle which were to follow, the species seems to have domesticated itself rather than been chosen for any particular use by man, unless it was used for meat. It was probably tolerated because of its companionship and hunting ability. Later additions to man's livestock were horses, chickens, and the long-suffering donkey or ass. Among other domestications, the services of the camel were invaluable in man's conquest of the desert, and the elephant – though never truly domesticated – became a beast of war.

Indian elephants are often richly decorated with jewels and painting and are ridden by princes in festivals and processions. They are a symbol of power. Other elephants are put to more practical tasks. Although they require a long training and consume huge quantities of food, the three thousand or so working elephants in Asia earn their keep by the excellent work they do in the jungle, ploughing and logging. African elephants are much less used and are sometimes said to be less easily tamed. Thirty-seven beasts were used by Hannibal in his crossing of the Pyrenees in 218 B.C., but he found them expensive to keep and losses in battle were hard to replace. Nowadays heavy machinery carries out the tasks they might have performed in Africa.

Many other species have been kept or domesticated, for example as pest-killers (cats and ferrets), fur-bearers (chinchillas) and meat-producers (rabbits and scores of others). Some have had only passing popularity. The most adventurous experimenters in this field were the Old Kingdom Egyptians of about 2500 B.C., whose art shows, among other creatures, captive hyaenas being fattened up for the slaughter, and several species of captive antelope. They also kept pelicans for eggs. Not all of these animals were domesticated, however; some were merely tamed. The Russians have attempted the taming of the European elk (called the moose in North America) in this century, for purposes of riding and driving. It can be kept on farms and milked like a cow, and needs only bark and twigs to eat. Apart from the rather aggressive reindeer which, like its owners, has retained its nomadic habits, the elk is the only deer which has been domesticated. In the horse family, only the zebras and the Asiatic wild ass remain undomesticated, though at the end of the nineteenth century some were used experimentally for drawing carts and wagons. Their lack of stamina and the advent of the motor car soon put an end to their career as draught animals. However, a hybrid animal called the zebroid, offspring of a horse and a zebra, is still used as a pack animal on safaris in Kenya. A more unusual substitute for the horse was the ostrich, used by the Romans for pulling carts and riding.

Africa looks to the future

Although there have been many experiments, there have been few large-scale attempts at domesticating new types of animals since New Stone Age man settled down with his sheep, cattle and horses. When he colonised new areas he took the same animals with him, even though they were not always the best suited to the climate and other conditions. Domestic cattle introduced into central and southern Africa, for example, do not make the best use of the available grazing, are less used to water shortage than the native animals, and are fatally susceptible to a disease called nagana which is spread by the tsetse fly. Yet huge herds of antelope and other game roam the savannah, sharing out the grazing in a much more economical way than do cattle, immune to many of their diseases and representing a rich source of meat. The 1950's saw some extensive investigation of the possibility of controlled cropping of wild game, but it now looks as though the actual domestication of some species is a more practical proposition.

The ideal species for domestication should be sociable, docile, adaptable. The African buffalo is too savage and the giraffe, though its meat tastes like beef, is difficult to confine behind fences. It seems that the best hope for the future lies with the antelopes. Gazelle flesh was certainly eaten by the Romans, and the Egyptians kept various other antelopes for their leather, horn and meat. Today, blesbok are successfully ranched in South Africa. But the most promising newcomer on the domestic scene is the largest of the antelopes, the eland, which turns out to be an excellent producer of milk and meat, and is easily tamed. There are herds in Rhodesia which are allowed to run free during the day and are coralled at night, but the

The beautiful white horses of the Camargue in France are allowed to run semi-wild with the black bulls of the same area. As a result they have an understanding of the terrain and of the bulls which is invaluable to the men who herd them.

Far from home: the herd of eland at Askaniya Nova in the Ukraine is descended from African stock brought over in the last century and now acclimatised to life on the Russian steppes.

most famous herd of domesticated eland is at Askaniya Nova in Russia's Ukraine province. Introduced there in 1896, the animals have become acclimatised to life on the steppes and can be milked and even ridden.

The story of Askaniya Nova, now a centre for agricultural research, is an outstanding example of the power of man to control the destiny of his fellow creatures, to set the lives of animals upon a course determined by him. The centre started life as a German sheep-farming colony in 1830, and after varying fortunes fell in the late nineteenth century into the hands of one Friedrich Falz-Fein. He was a man with a vision: to create a wonderful animal sanctuary in the steppes, a haven of peace where no bird was caged and all species could roam at will. And he succeeded. He sank new wells for water, planted trees, and established an oasis of greenery which he filled with wild creatures from all over the world – ostriches, emus, doves, kangaroos, llamas, deer, zebra, American bison, and innumerable others. Even today there are still some of the few surviving pure-bred Asiatic wild horses there.

Keeping animals in captivity and in sanctuaries, protecting them from the hunter's gun, eliminating the struggle from their lives – so much of peace in the animal world is in the hands of man. His power over nature can be put to good use in protecting the future of species which have become rare. With competition for food and living quarters at a minimum, and hunting banned, an endangered species has a much-improved chance of recovering in numbers. In general, wild animals kept in captivity have an easy life and live longer than their free relatives, and their confinement may often be justified by their value to man in research, education and even entertainment. But few people would deny that, whenever possible, animals should be left in the natural, wild state. Though their lives seem to be seldom free from conflict, this is the normal pattern of their existence and should be preserved.

Man has gained mastery over the wild. Now he can choose whether to hold all animal life in bondage, or to give back that most prized of gifts, freedom.

Indexes

Numbers in italics indicate illustrations

Animal index

Subject index

For permission to use published/copyright material we would like to thank the following:

The Oxford University Press and The Society of Jesus for the extract from Gerard Manley Hopkins' poem 'Pied Beauty'; The Clarendon Press for an extract from 'The Flycatchers' by Robert Bridges; Messrs Frederick Warne & Co for the extract from Beatrix Potter's 'The Tale of Peter Rabbit'.

Photo Credits

Cover – Albert Visage: Jacana
1 – Animal Photography
2 – Yves Lanceau
7 – Jane Burton: B Coleman
9 – Arthur Christiansen
10 – Sven Gillsäter: Tiofoto
13 – MWF Tweedie: NHPA
14,15 – Fred Bruemmer
16T – Root/Okapia
C – Arthur Christiansen
B – Philippa Scott
17 – Peter Ward
19 TL – GS Giacomelli
TR, C,BL – PH Ward
BR – Stephen Dalton: NHPA
20,21 – PH Ward
22 – MWF Tweedie: NHPA
24 – Zool Soc London
25 – Röhdich: Bavaria
27 – Toni Angermayer
29 – C Vienne: Jacana
30 – Jane Burton: B Coleman
31 – P Morris
33T – Jane Burton: B Coleman
B – HR Bustard
36 – roebild
37 – JH Moon
38T – Joe B Blossom: NHPA
B – H Hansen
39T – H Schrempp
B – R Apfelbach
40 – M Thurston
42 – Arthur Christiansen
43 – Andre Fatras
44 – Toni Angermayer
45T – Russ Kinne: B Coleman
B – M Thurston
47 – L Lee Rue: B Coleman
49T – Okapia
B – Keystone
50 – Toni Angermayer
51 – RD Martin
53 – RS Virdee
54 – Wisconsin Regional Primate Research Center
55 – Okapia
56 – Usclat: Jacana
57 – roebild
58,59 – G Kinns
60T – W Vandivert
B – H Schrempp
61 – Paul Steinemann: Basel Zoo
62TL – Zool Soc London
TR – DJ Chivers
B – Lynwood M Chace
63 – Max Lenz: WWF
64 – Popperfoto
66 – Sven Gillsäter: Tiofoto
67 – Georg Nystrand
68 – Okapia
69 – WWF
70,71 – Alan Band Associates
72L – P Morris
R – G Kinns: AFA
73 – Toni Angermayer
75 – G Rüppell
77 – Anthony Bannister: NHPA
78,79 – L Lee Rue: B Coleman
80 – H Schrempp
81T – GE Hyde
B – J Allan Cash
82 – Bavaria
84 – Brian Hawkes: NHPA
85T – H Schünemann: Bavaria
B – Roy A Harris & KR Duff
87T – Des Bartlett: B Coleman
B – Zool Soc London
89 – Stephen Dalton: NHPA
90 – South African Tourist Corporation
91 – Popperfoto
94 – roebild
95 – Imre von Boroviczeny: Bavaria
97 – Peter Hill
98,99T – Okapia
98,99B – Alan Band Associates
100 – J Robert: Jacana
101 – JB Free
103 – Barnabys
104T – Bernard Rebouveau: Jacana
B – Jane Burton: B Coleman
107T,B – Colin G Butler
C – JB Free
108 – AN Warren
109 – Photographic Library Australia
110 – Peter Hill
111 – Barnabys
113 – Simon Trevor: B Coleman
114T – John Warham
B – Arthur Christiansen
115 – Heather Angel
116 – W Wales
117T – John Tashjian at Steinhart Aquarium
B – LE Perkins
118T – KB Newman
B – Jean-Philippe Varin: Jacana
119 – WM Stephens
123 – Graham Pizzey: B Coleman
124 – Douglas Fisher
126 – Okapia
128T – Peter Hill
B – Stephen Dalton: NHPA
129 – Argus Africa News Serivce
131 – Photographic Library of Australia
133 – Derek Bayes/Observer
134 – Douglas Fisher
135 – Kaj Boldt
136 – Barnabys
138 – Popperfoto
139 – L Lee Rue
140 – Peter Hill
141 – EO Hoppe
143T – Manchester City Art Galleries
B – Barnabys
144T – from FE Zeuner's 'A History of Domestic Animals' Hutchinson 1963
B – RC Bigalke
145 – Okapia
146,147 – HW Silvester: B Coleman
148 – Okapia